THEOLOGICAL ANALYSES OF THE CLINICAL ENCOUNTER

Theology and Medicine

VOLUME 3

The titles published in this series are listed at the end of this volume.

THEOLOGICAL ANALYSES
OF THE CLINICAL ENCOUNTER

Edited by

GERALD P. MCKENNY

Rice University, Houston, Texas, U.S.A.

and

JONATHAN R. SANDE

Mayo Clinic, Rochester, Minnesota, U.S.A.

Kluwer Academic Publishers

Dordrecht / Boston / London

Library of Congress Cataloging-in-Publication Data

```
Theological analyses of the clinical encounter / edited by Gerald P.
  McKenny and Jonathan R. Sande.
       p.    cm. -- (Theology and medicine ; v. 3)
   Includes bibliographical references and index.
   ISBN 0-7923-2362-9 (alk. paper)
   1. Medicine--Religious aspects.  2. Healing--Religious aspects.
  3. Physician and patient.  4. Covenants--Religious aspects.
  I. McKenny, Gerald P.  II. Sande, Jonathan R.  III. Series.
  BL65.M4T45  1993
  261.5'61--dc20                                             93-17958
```

ISBN 0-7923-2362-9

Published by Kluwer Academic Publishers,
P.O. Box 17, 3300 AA Dordrecht, The Netherlands.

Kluwer Academic Publishers incorporates
the publishing programmes of
D. Reidel, Martinus Nijhoff, Dr W. Junk and MTP Press.

Sold and distributed in the U.S.A. and Canada
by Kluwer Academic Publishers,
101 Philip Drive, Norwell, MA 02061, U.S.A.

In all other countries, sold and distributed
by Kluwer Academic Publishers Group,
P.O. Box 322, 3300 AH Dordrecht, The Netherlands.

Printed on acid-free paper

Printed in the Netherlands

TABLE OF CONTENTS

GERALD P. MCKENNY

THEOLOGY, ETHICS, AND THE CLINICAL ENCOUNTER

An introductory essay to a volume of diverse papers assumes two duties toward its readers. The first is to provide a general thematic unity for the essays. The second is to summarize the essays themselves. I attempt to fulfil both of these duties in the current essay. The first part explores the tasks of theology in health care ethics by summarizing current options and their representation in the present volume. It also raises the question of the distinctiveness of ethics in the clinical setting. The second part locates the essays in current debates in health care ethics by sketching the key issues at stake in some of these debates and then summarizing the essays themselves. Hence the present essay is not merely an introduction to the volume but a map of the territory in which it resides.

THEOLOGY AND CLINICAL ETHICS

The question of the proper role of theological claims in health care ethics is a subject of ongoing debate. One place to begin a survey of the options is with an influential debate. In the conclusion to his 1975 Pere Marquette Lecture, James M. Gustafson wrote that

For most persons involved in medical care and practice, the contribution of theology is likely to be of minimal importance, for the moral principles and values can be justified without reference to God, and the attitudes that religious belief grounds can be grounded in other ways. From the standpoint of immediate practicality, the contribution of theology is not great, either in its extent or its importance ([4], p. 94).

Two years later his former student, Stanley Hauerwas, countered:

To be sure, Christians may have common moral convictions with non-Christians, but it seems unwise to separate a moral conviction from the story that forms its context of interpretation. Moreover, a stance such as Gustafson's would seem to assume that medicine as it is currently formed is the way it ought to be. In this respect, we at least want to leave open the possibility of a more reformist if not radical stance ([5], p. 203n.).

The debate between Gustafson and Hauerwas reflects a wider debate of the 1970s and 1980s between those theologians who, like Schubert Ogden [10] and David Tracy [14], sought to relate theological claims to the claims of other realms of inquiry and practice, and those who, like George Lindbeck [9], sought to describe the language and practices of particular communities of faith. But as Gustafson's subsequent work made clear, affirmations of the compatibility of theological and other claims do not necessarily entail inattention to

G.P. McKenny and J.R. Sande (eds.), Theological Analyses of the Clinical Encounter, vii–xx.
© 1994 *Kluwer Academic Publishers. Printed in the Netherlands.*

the particular contexts of those theological claims or lack of criticism of prevailing practices. Gustafson's remark is directed against those who would claim either that theology can ground the whole enterprise of medical ethics or that it can supply principles and attitudes essential to the ethical practice of medicine that can not be otherwise derived. It does not rule out a role of reminding medicine of its ethical ideals, of contributing to disputes within the field, or of taking stands on issues. Above all, it does not mean that any other discipline is any more capable of grounding the enterprise; theologians may constitute one voice among others depending on the strength of their arguments. But it does seem to assume a general convergence on principles and attitudes that is recognized by all participants in the debate – a convergence that is becoming increasingly more difficult to define. It also seems to assume that disagreements will occur largely within a realm already circumscribed by a recognized set of principles and attitudes. In these last two respects Hauerwas may seem to have the upper hand.

But there is a tension between Hauerwas's appeal to a particular community and his reformist or radical stance. His turn to particularity does not entail a sectarianism in which criticisms have no public validity, as is often charged. But while his always insightful and often compelling criticisms of some characteristics of technological health care appeal to a wider public, his solutions usually appeal to a community simplistically characterized as having been formed by Christian convictions embedded in stories.

A careful reader of the essays that comprise the present volume will hear echoes of this debate, though without the confidence of Gustafson in fundamental agreement or the confidence of Hauerwas in the existence of the story-formed community. Should theology identify points of convergence between particular claims of religious traditions and the claims of secular traditions in order to arrive at a public consensus, as many have argued in recent years? Or should theology bring to attention specifically religious issues that are inseparable from many of the stories of both practitioners and patients but which are usually obscured by the reigning methods of moral inquiry, as Heitman, Lauritzen, and Lebacqz suggest? Should theology seek to offer a more adequate account of norms, rules, principles, or virtues that govern the clinical encounter, as Glaser, Hamel, May, Pellegrino and Thomasma, and Sulmasy suggest? Or should theology participate in a much more critical role, undermining false assumptions deeply imbedded in the discourse and social practices of the clinical setting, as DuBose argues and as a theological appropriation of Churchill and Churchill would presumably argue? Or finally, should theology seek to elicit and describe a moral encounter that is phenomenologically prior to all our moral theories, as a theological appropriation of Zaner might do?

These questions, respectively, bring current issues of universalism and particularism, reconstruction and deconstruction, and alterity to the interface between theology and the clinical setting. Few would argue that the present essays provide conclusive answers to these questions. But they do make the case for placing theology on the cutting edge of debates about ethics in the clinical setting.

Clinical ethics may be even more difficult than theology to characterize. Proponents of clinical ethics such as Albert Jonsen [6] and Mark Siegler [13] have sought to distinguish their methods from those of theoretical bioethics, but their efforts are not convincing. An ethic of principles would seem to be more theoretical than clinical, but Richard Zaner [15] has shown that there is no significant difference between the methods of moral decision making in Jonsen, Siegler, and Winslade [7] on the one hand and Beauchamp and Childress [2] on the other, although the former advocate a clinical approach while the latter advocate a principles approach. Similarly, a casuistic ethic would seem to be more clinical than theoretical, but John Arras [1] has shown that casuistry as recommended by Jonsen and Toulmin [8] can not avoid theoretical questions faced by other approaches. Nor is there any easy methodological distinction to be drawn between clinical ethics and policy ethics. Virtue ethics would seem to favor the clinic, but who can deny the relevance of virtues such as courage and discernment to policy issues? A Rawlsian method of reaching an overlapping consensus amid diverse view-points seems best suited for the policy arena, but its method also seems applicable to clinical cases. Finally, both casuistry and quantitative schemes for resolving issues of allocation are being proposed for both the clinic and the policy sphere. It remains to be seen whether the interesting work being done in this volume and elsewhere on narrative and rhetorical forms of moral reasoning will provide a model of moral reasoning uniquely suited to the clinical setting.

This survey indicates that clinical ethics is distinguished not by method but by content. Most of the authors in this volume raise questions about standard approaches to ethics not by proposing a unique method but by pointing to specific characteristics of the clinical setting and their implications for ethical reflection. While it may be difficult to draw methodological boundaries between clinical ethics and its neighbors, it is certainly possible to describe the differences in terrain and the objects that populate the diverse realms. Appeals to actual clinical practice, to concrete features of cases, and to the spatial and temporal configurations of relations between persons, events, and institutions demarcate the "clinical" in this volume.

COVENANT, PRINCIPLES, AND BEYOND

The essays that comprise this volume fall into three distinct forms of moral discourse in health care ethics. The first concerns a religio-moral ideal, that of covenant, which continues to exert some influence. Moral ideals can give rise to obligations or can be expressed in virtues. The covenant model has taken both forms, and thus straddles contemporary debates in ethics.

The relation between specifically religious claims and moral requirements is obviously a problem any covenant model faces. Not all persons would share Paul Ramsey's optimistic conclusion that the religious content of covenant faithfulness is the inner meaning of which the norms of justice are the outward expression ([12], p. xii). Perhaps this is why many theologians working in health care ethics have been more inclined toward the second approach, which is based on principles which are allegedly logically independent of controversial theological claims. Yet principles themselves are vulnerable on these very grounds to criticisms that they abstract from concrete moral commitments aside from whatever theoretical problems they might incur (see below).

Covenant ideals and principles alike presuppose features of moral reasoning and moral agency that have now become controversial. Their tendencies to exclude or minimize some capacities of moral agency (such as emotion, commitment, and sensitivity to particulars) and to overestimate other capacities (such as transparency of language and intention, transcendence of class and gender bias, and objectivity) are being exposed by research carried out under the auspices of many disciplines and approaches.

It may appear, therefore, that the three sections of this volume represent a temporal progression from covenant (past) to principles (present) to something beyond principles (future). But things are not this clear. Criticisms and reformulations of the covenant model draw upon materials that might seem to belong in the third section. And arguments in the third section contain echoes of premodern ideas. The interface of theology and the clinical encounter appears destined to this mingling of tenses for the foreseeable future. But it is certainly not alone in this respect.

1. The Medical Covenant Past, Present, and Future

The discussion of the contribution of theology to medical ethics is usually treated as part of the explosion of interest in medical ethics during the past twenty-five years. In reality, of course, theological interest in medical practice like medical ethics in general has occurred for centuries in the west. But just as the current movement in medical ethics was precipitated by the simultaneous collapse of traditional patterns of moral authority and the rise of new technology, so the recent discussion of the role of theology in medical ethics was fueled by the need to address these changed circumstances. The time-honored traditions of casuistry as practiced especially in the Jewish and Catholic traditions seemed ill-equipped to handle the flight from authority and the new technology alike.

In this environment, the notion of covenant seemed the ideal trope for several reasons. First, it emphasized the interpersonal character of the relation between doctor and patient when technology and bureaucratization seemed to be undermining the traditional ideal of the therapeutic relationship. Second, covenants, while interpersonal, may be expressed in mutual rights and duties. This answered the need for specific publicly accepted rules to replace the discredited medical paternalism and religious casuistry. Third, the notion of covenant, rooted in the Hebrew Bible, well established in Protestant theology, and its general contours increasingly congenial to post-Vatican II Catholic theology, seemed to be ecumenical. Finally, and perhaps most importantly for theologians eager to secure their place in a rapidly growing yet apparently secular field, the concrete guides to conduct that follow from a covenant model seemed to converge with secular thought. After discussing the grounding of his ideas in the biblical covenant ideal, Paul Ramsey confidently remarks: "in the midst of any of these urgent human problems, an ethicist finds that he has been joined – whether in agreement or with some disagreement, by men of various persuasions, often quite different ones" ([12], p. xii).

All four of these features of the covenant model are currently under attack. The covenant model has not been able effectively to counter the problems of bureaucratization, power, and distrust in the doctor-patient relationship. Except in the hands of William F. May, its emphasis on rights and duties now appears to have contributed to the neglect of virtue and character in medicine. In an era of pluralism and interreligious dialogue it appears to be not only narrowly Jewish and Christian, but even more narrowly Protestant in its theological claims. And finally, the alleged convergence in practice with secular thought turns out to have been due to the ease with which covenant, its theological claims muted, assimilates into contract.

The essays in the first section of this volume address the notion of covenant and assess its historical place, its limitations, and its prospects for the future. Ron Hamel begins with a survey of the covenant model in the work of Paul Ramsey, Robert Veatch, and William F. May. He probes the perceived shortcomings of contemporary health care practice that led each to adopt a covenant model: a concern for the vulnerable in the face of bureaucratic and utilitarian tendencies (Ramsey), a concern for preservation of basic moral norms among strangers and in face of the power of medicine (Veatch), and a concern with the evasion and denial of death that puts physicians at a moral distance from their patients (May).

The careful reader can also find in Hamel's expositions three approaches to theology in medical ethics: Ramsey's convergence, in which theology provides the internal and deep meaning of which secular norms are the external condition or embodiment, Veatch's medical contract as an attempt to preserve the theological meaning of covenant without paying the particularistic consequences, and May's use of covenant, which can not fulfil the requirements of a universal morality but is not merely the conviction of a particular community, to explore features of medical practice that are ignored in secular accounts. Hamel is equally sensitive to the accomplishments of the covenant model (it allows for a breadth and depth lacking in other models), its limitations (it is too narrowly focused on the relation between doctor and patient), its ambiguities (is it individualistic or communal in nature?), and its possibilities (its capacity for focusing on essential features of the clinical encounter that are easily overlooked).

William F. May's essay relates his earlier work on covenant to his current work on the marks and virtues of the professional. He accomplishes this task by means of a twofold critique of obligation theories in ethics. First, he admits that a covenant model must involve obligations but decries the "uncoupling of narrative and moral principle" that accompanies attempts to reduce a covenant ethic to a theory of obligation. Second, he argues that obligation theories fail to account for how the fact that one is a professional affects what one brings to the classical dilemmas and their resolution by principles. May identifies three marks of the professional and discusses their corresponding virtues of prudence, fidelity, public spiritedness, and their attendant virtues. He engages biblical meanings of covenant by tracing these virtues to gratitude – "the sense that one inexhaustively receives" – which presupposes or signifies a more transcendent source of giving than the sum of what is received from others. Recognition of gratitude in turn leads to hope based on promise, another feature of biblical covenants. While some theologians and philosophers will wish in vain for an rational argument to justify these claims, others will appreciate the way May's "musement" (to use Charles Sanders Peirce's term) on gratitude and hope gestures toward sensibilities or convictions that are often ignored in recent theories of virtue.

Historically, the covenant model often sought to account for and to preserve the trust that was considered the most fundamental condition of the clinical encounter. But does the covenant model rest upon a mistake? Are we in an era in which the fiduciary expectations patients have of health professionals have been disappointed and in which distrust rather than trust increasingly becomes the means of social control and the medium of exchange in the clinical encounter? And is the ethic of contract proposed by proponents of a secular bioethic a late effort to ensure trust that has already yielded to suspicion of professionals and their power?

Edwin R. DuBose raises these questions in an essay that issues a sharp challenge to advocates of a covenant model. If indeed distrust has become the new form taken by the clinical encounter then DuBose finds two interpretations of the situation. The first interpretation is that public trust is declining but is fixable by dealing with the commitment and character of professionals and by addressing the conflicting interests dividing them from patients. Trust can be restored if physicians are of good commitment and character. The other interpretation is that the very fiduciary claims themselves – the claims to present facts, act in the patient's best interest, and be committed to the patient in the form of professional virtues – conceal and support the very factors that undermine trust. Unlike May, DuBose explores the second interpretation, which few bioethicists have been willing to countenance. He indicates the complicity of covenant models in the concealing and supporting, insofar as they have underwritten yearnings for a transparent relationship and a fulfilment of all that the patient lacks – goods that medicine can not possibly deliver. But DuBose does not wish to jettison the covenant model entirely: he closes with a plea for more limited and particularized covenants.

2. *Principles in Revision*

Competing with covenant models and their secular analogues have been various versions of an ethic of principles. The work of Tom Beauchamp and James Childress, and Albert Jonsen, Mark Siegler, and William Winslade represent two forms of this often misunderstood approach which at its best aims at a careful integration of theory and casuistry. The approach denies from the outset any theological grounds for its validity. Despite this, theologians have not hesitated to make extensive and sometimes uncritical use of principles such as autonomy, non-maleficence, beneficence, and justice.

Principles approaches in bioethics have endured ongoing criticism from virtue theorists and casuists, but recently they have been sharply challenged by K. Danner Clouser and Bernard Gert for shortcomings and inconsistencies in their own claims [3]. Since principles such as autonomy may appeal to two or more incompatible theories for justification, the coherence of a principles approach is challenged. Another argument is that the principles themselves do not constitute action guides; hence rules must somehow be derived from them, but it is not clear how this is to be done. Yet another criticism is that principles themselves can not resolve conflicts between two or more principles, say autonomy and beneficence. A final criticism is that the content of the principles themselves is eclectic and that efforts to delimit or order the content of the principle – for example the priorities among not harming, preventing harm, removing harm, doing good in the principle of beneficence – are bound to be arbitrary unless a more rigorous theoretical derivation and justification of principles is undertaken.

For these reasons, claims to justify such principles on rational grounds accessible to all persons have fallen upon hard times. But the popularity of principles in health care ethics probably can not be explained simply by an intellectual laziness that avoids theoretical rigor. Is it possible that principles serve as a conceptual shorthand for an overlapping consensus or a convergence of intuitions that command more agreement than the theories from which the content of the principles is, admittedly, rather eclectically drawn? Are orderings of content within and among principles done according to the demands of particular cases rather than from the gaze of theory?

The question for our purposes is whether theology has anything to contribute to this debate which until now has been carried out between moral philosophers and casuists. But by far the principle that has occasioned the most controversy in health care ethics in recent years is the principle of autonomy. All four of the essays in this section address limitations of autonomy and seek to locate it in a wider range of moral values.

Currently the primacy of autonomy is being challenged on several grounds. First, clinicians and ethicists suggest that the practice of medicine and other health professions is based on a claim to beneficence on the part of the practitioner and relies upon mutual trust, and that these features imply a more relational model of the clinical encounter. David Thomasma and Edmund

Pellegrino now address the theological claims implied in their effort in recent years to restore beneficence and trust to medicine.

The line of inquiry into beneficence opened up by Pellegrino and Thomasma [10] is also generating further inquiry. Two of the key targets of the modern medical ethics movement were paternalism and bureaucracy: autonomy was often justified on the grounds that patients needed to protect themselves against the unwarranted authority of physicians and against the dehumanizing character of institutions. Today the attack on authority is itself being attacked for sharing a common discredited modernist conception of authority while new insights into the power and effects of institutions are emerging from social scientific research. The clinical encounter can not escape questions about the moral status of the authority relations that comprise it nor the institutional loci in which it dwells. Daniel P. Sulmasy and John Glaser, respectively, address break new ground on questions of authority and institutions in health care ethics.

Finally, autonomy-based theories assume a formal equality of moral agents in relevant respects that abstracts from what many consider the most pervasive feature of illness: vulnerability. But once the imbalance is recognized, attempts to subsume questions of power under the rubric of autonomy can be accused of obscuring the role different forms of power actually play in the clinical setting. Karen A. Lebacqz examines issues of power and powerlessness in the clinical setting.

Thomasma and Pellegrino ask whether a Christian perspective provides a corrective to the absolutization of autonomy in medical ethics. More specifically they ask whether a theological perspective can place autonomy within a context of a trust relationship for both physician and patient. Thomasma and Pellegrino proceed by describing the role respect for persons has within the context of beneficence. Patient autonomy becomes an obligation of beneficence or charity on the part of the physician, but at the same time respect for the autonomy of the professional or of the institution becomes an obligation of beneficence on the part of the patient. Moreover, in both cases a theological perspective raises questions about the way autonomy is used. Hence duties qualify freedom, though the overriding of freedom by another person in the name of a duty is ruled out.

Thomasma and Pellegrino argue that theological reflection reveals a depth and substance to the idea of autonomy within beneficence from a Christian perspective that is lacking in secular accounts. The recognition of duties that qualify freedom, the tendency of beneficence to violate the boundaries set for it in philosophical reflection, the special concern for the vulnerable, and the mutuality of respect for freedom are all supported by a theological perspective. The themes of vulnerability and mutuality of respect give rise to reflection on the inevitability of trust, which the authors identify as the most fundamental characteristic of the clinical encounter. This leads to reflection on the way faith both enables and qualifies trust between physicians and patients. Fidelity to trust emerges as a further obligation of beneficence or charity.

Sulmasy opposes the modern view of authority derived from Thomas Hobbes and John Locke that places authority in opposition to or in conflict with autonomy. Authority in this model is possessed by one party over another in the form of expertise or control. Sulmasy argues for another kind of authority, the authority "to act in the freedom granted one by someone else." He turns to the New Testament for a version of authority as *exousia*. *Exousia* is not possessed by either party but transcends both. It consists in moral warrant rather than in control or expertise. Ultimately it is granted by God through the practice of medicine in which doctor and patient join in a shared telos.

Sulmasy's essay is a strong contribution to current efforts to find a third way between autonomy and paternalism in medical ethics. More broadly it falls in line with current efforts in the humanities to critically recover features of tradition that were obscured under Enlightenment assumptions. Ethically, the argument rests on an effort to deliver the notion of authority from anxieties about control and dependence to the moral sphere while avoiding those features of authority that contribute to control and dependence. Theologically, it rests on "mutual recognition by both the healer and the healed of the ultimate source of the power to heal and the ultimate source of the warrant to heal," though Sulmasy also argues that this claim can be translated into the language of a practice and its telos.

Glaser's argument for an ethic that recognizes the institutional factor in the clinical encounter rests upon two claims. The first is that from a Christian perspective beneficence is the central principle of health care ethics. The second is that moral conflict, the choice between conflicting goods, is not an occasional occurrence but the very fabric of the moral life. As a result, while benevolence wills good for all, beneficence is "love-within-limits."

Moral conflict is inevitable in health care because all interactions occur in a relationally complex network involving individuals and their goods, organizations and their goods, and society and its good. Human beings are always "persons-in-relationships" and organizations "are woven into a matrix of society." For Glaser this complex network of values demands that moral judgments be made in a community of concern that is broad enough to address the complexity of the issue at hand. The composition of the community should reflect the value relations that are at stake in any given issue. Glaser also discusses various types of conflict between the individual, organizational, and societal levels in health care.

Lebacqz takes issue with tendencies to subordinate questions of power to those of autonomy. When this happens, empowerment is understood only in terms of enhancing or restoring autonomy. But Lebacqz, drawing upon the lexicon of principles, claims that empowerment belongs under the heading of justice as well as that of autonomy. Guided by the conviction that instances of disempowerment provide the epistemological key to empowerment, Lebacqz discusses two case stories that represent disempowerment among the relatively powerful and the relatively powerless, respectively. She finds several forms of disempowerment: physical; moral or spiritual; and social, economic, cultural, and gender related forms.

Lebacqz keeps each form of disempowerment in focus without ignoring any or attempting to reduce one to another. She also shows how being relatively powerful or powerless determines which forms of disempowerment are experienced, how they are experienced, and what empowerment would mean. The result is a strong case for empowerment as a new topic in medical ethics, mediating between but also transcending the standard concerns of autonomy and justice.

The essays in the second section decry the incapacity of the autonomy model and its assumptions to account for these features of the clinical encounter. They constitute an interesting test case for drawing a distinction between theoretical and clinical ethics since each author finds features in the clinical setting that challenge the armchair assumptions of principles, yet none argues that principles should be abandoned. Equally important, they offer alternative accounts of principles which make specific claims for the relevance of theology to the task of revising principles. These claims range from arguments that such principles are more adequately grounded in theological than in secular assumptions to the suggestion that such principles need to take more account of religious aspects of illness.

3. *Beyond Principles*

Narrative, rhetoric, otherness, pluralism. . . . These terms form an important part of the lexicon of "postmodern" discourse in the humanities and social sciences but they are only beginning to work their way into the discourse of medical ethics. Discontent with Enlightenment ideals is sometimes exaggerated and the discontented frequently appeal unwittingly to Enlightenment assumptions. But since modern ethical theories did presuppose moral agents similar in relevant respects, firmly in control of their actions, and able to debate as free equals, while the actualization of these conditions in concrete settings was endlessly deferred, discontent seems justifiable.

Some readers may be tempted to conclude before reading this section that these concerns are trendy and have little to do with the actual practice of health care. But consider a relatively simple controversy in which a doctor presents a case involving a patient who refuses lifesaving surgery and another doctor or a nurse has a distinctly different impression that the patient is crying out for help. Who is right? It is futile to ask what is "really" happening: both believe they know what is really happening and there is no real event itself to which a neutral observer could appeal in order to resolve the problem. Appeals to rules, principles, or ideals are similarly futile – a description of the case must be established before rules or principles can be applied, but this is exactly the point of controversy. The "case" is inseparable from the way it has been represented. Each side tells a certain story that organizes the flux of experience into an event that can be read as a case before an audience that never saw the patient. Moreover, the categories and perspectives each participant brings to the encounter determines what will be selected out of the flux and how it will be ordered. No one doubts that a person trained in medicine will "see" a different event than a lay person, and only by imaginative efforts of abstraction and translation are we able to speak of the "same" event. Principles, rules, ideals, and stories are various ways of creating events out of the flux of the encounter – of telling us what to look for and persuading us of what is there. And of course, our convictions about what is there will also determine what principles, etc. we find convincing.

Viewed from this perspective, the problem with health care ethics has been an underlying assumption that there is a single true interpretation of a case and that rules or principles, which are similarly transparent, need only be applied according to rationally defensible procedures. Without necessarily arguing that we have only representations, these new approaches argue that one primary task of moral reflection is to subject representation itself to scrutiny. The first step in this type of moral reflection is necessarily negative. The lines of criticism pursued in these essays undermine some of the most fundamental conditions of previous types of health care ethics. Assumptions that moral reason ought to be impartial and abstracted from narrative contexts, that moral agents are in command of a single line of moral reasoning or a single narrative, and that medical morality is grounded in a free encounter of independent equals are all interrogated and found wanting. This attack upon some of our most cherished ideas about morality is bound to be unsettling to some and is not to be taken lightly by anyone. But like much postmodern criticism, the intent in each of these essays is constructive; the end of the moral discourses of modernity is viewed as an opportunity for a more adequate medical morality. The very task itself is premised upon profound moral concerns, many of which also animate modern moral theories but which often can not come to expression in those theories.

Pluralism, another concern of postmodern thinkers, is scarcely a new topic in medical ethics. Most theoretical and much practical work in medical ethics presupposes diversity: this is one rationale for the emphasis on respect for self-determination and the appeal to negotiation between health professionals and patients. However, the assumptions of autonomy, self-determination, and the independence of rational individuals which usually accompany efforts to deal with pluralism are called into question by the need to deal appropriately with patients whose beliefs about health and illness and expectations of healers conflict with such assumptions. This calls for both greater attention to pluralism at it appears in clinical settings and alternative ways of dealing with what is found there. Studies of religious and cultural beliefs and practices with regard to health care abound in the literature of medical anthropology. Curiously, however, there has been scant attention to cultural difference as a challenge for medical ethics in the clinical setting.

Paul Lauritzen explores the eclipse of emotions in modern moral theories and its effects upon clinical discernment. He argues that the emphasis on impartiality in moral reasoning has led to the eclipse of the emotions not only as a source of moral insight but also as a capacity of moral discernment. Lauritzen indicates the precise sense in which impartiality involves abstraction and how this affects emotional factors as material for moral judgment. He then draws upon two clinical cases to show how the eclipse of the emotional responses of the parties involved obscured factors in the cases on which moral judgments hinge.

Lauritzen does not simply ask clinical decision makers to incorporate their own emotions into moral judgments or to be more aware of emotional factors in cases. Rather he proposes narrative forms of moral reasoning that develop and cultivate emotional capacities and skills of discernment which in turn enable decision makers to recognize and be moved by crucial concrete details of patients' stories that are missed from an impartial perspective. Such details will in many cases involve religious beliefs. Unfortunately, however, the translation of religious beliefs into universally accepted conclusions, which occurs under the assumption that conclusions are all that matter, distinguishes seeing from feeling in patients' religious beliefs and may therefore have the ironic effect of rendering religious responses of patients more opaque by seeking to render them more intelligible.

Sandra Churchill and Larry Churchill agree with the turn to narrative but argue that in some forms it is subject to the same criticisms as a principles approach. Their recommended alternative is a rhetorical approach. Their central claim is that a "case" in a clinical setting is never simply "about" a matter concerning which a decision must be made, but is also the way in which it represents what it is "about." Its mode of representation inevitably highlights certain features and plays down or excludes others. Narrative ethics is in danger of substituting a tyranny of representation with a single point of view and a single linear sequence for the tyranny of principles with its single line of argumentation and vertical movement from principles to cases. Both assume that there is only one line of reasoning to be applied or one story to be told about a case. Deductive systems of principles proceed "as if agents were in an administrative posture over themselves and their actions." Narrative approaches overcome this problem but assume a linear sequence of agency and imply the sufficiency of the narrator's point of view.

Discussing a well known case involving physician assisted suicide, Churchill and Churchill show how rhetorical analysis makes explicit the implicit selectivity in narrative representation. Rather than simply assuming that the agent or narrator is in full control of his or her actions and their meanings, the rhetorical approach reveals human agency as operating in a complex web involving both the field of action and the way it is represented.

While Churchill and Churchill question the fundamental assumptions of the reigning forms of moral reasoning, Richard Zaner turns to fundamental questions about what constitutes the clinical encounter as a moral encounter. Hence he is not immediately interested in the question, "what should I do?," but in the prior question, "who am I with this patient in this setting?"

Answering this question requires reflection on the spatial, temporal, textural, and bodily features that constitute the clinical encounter but which are abstracted from in standard moral argumentation.

Zaner proceeds by way of an extensive phenomenological description of a clinical case. He arrives at what he sees as the moral root of the clinical encounter: the awakening or evoking of a fundamental moral sense that is usually dormant and that emerges in an *ec-static* response to the other. Reflecting further on this *ec-stasis*, Zaner arrives at a description of the reflexive relations that constitute self and other in the clinical encounter. He concludes with an ethic of "being oneself by and as being with others," an ethic of compassion that combines care and trust. Zaner's method and conclusions will probably be rejected by those who believe the major task of ethics is to resolve disputes in ways that are clear and agreeable to all. But they will be of great interest to those who believe that part of the task of ethics is to bring moral responsiveness itself to awareness and to make agents more sensitive to the moral claims of concrete others.

Elizabeth Heitman explores the problem of pluralism in health care ethics and suggests a new model of the clinical encounter. After summarizing the extensive evidence for cultural diversity in forms and styles of communication in the clinical setting, language barriers, beliefs about the diagnosis and treatment of disease, and expectations about the relationships between healer, patient, family, and community, Heitman argues that proposed solutions to problems of pluralism presuppose particular western views and thus fail to address the problems faced by health professionals and patients from different cultural backgrounds. She also identifies clinical consequences of failing to deal with such diversity.

How can western doctors respect cultural differences while still maintaining commitments to the patient's health and to scientific medicine? Heitman finds in this question an interesting parallel to the chief question of interreligious dialogue: how can one affirm other religious traditions while remaining faithful to one's own? In both cases, she argues, the solution lies in dialogue as "an open-ended, ongoing process of mutual inquiry and self-discovery." She closes by showing how this model might help resolve several concrete problems that arise between western and non-western practitioners and patients in clinical settings.

CONCLUSION

Readers will have to judge for themselves what future theological reflection has in clinical settings and which arguments and strategies theologians ought to employ. Some of the essays, most of them authored by Catholic thinkers, make few concessions to secularism and pluralism and articulate a confidence that theological concerns can either converge with secular insights or accomplish tasks that secular approaches can not. But other essays indicate that theological reflection is in the same position as moral reflection, namely without a single vocabulary or set of problems that would delineate a secure place. Hence the attention given in many of these essays to clearing a space for theological concerns to occupy or discussing barriers to clinical appropriation of religious discourse. Still other essays do not venture theological claims at all.

Perhaps the essays reflect quite accurately the current status of theology in the clinical setting. For rather than proclaiming a correlation with principles and attitudes derived from other sources or describing the grammar of a stable community, theology seems destined to one of two fates: to recover particular traditions and identify ad hoc convergences with others or to join other perspectives in a more peripatetic, eclectic mode of inquiry. Perhaps only when theologians become comfortable with these alternatives will the unceasing (and unedifying) anxieties and energies gathered together in the task of clearing a space be devoted to more critical and constructive endeavors. The hope is that these essays will contribute to such endeavors in the area of clinical ethics.

Rice University
Houston, Texas
U.S.A.

BIBLIOGRAPHY

1. Arras, J.: 1991, 'Getting Down to Cases: The Revival of Casuistry in Bioethics', *Journal of Medicine and Philosophy* 16, 29–51.
2. Beauchamp, T., and Childress, J.: 1979, *Principles of Biomedical Ethics*, 3rd ed., Oxford University Press, New York, 1988).
3. Clouser, K. D., and Gert, B.: 1990, 'A Critique of Principlism', *Journal of Medicine and Philosophy* 15, 219–36.
4. Gustafson, J. M.: 1975, *The Contributions of Theology to Medical Ethics*, Marquette University Press, Marquette, Wisconsin.
5. Hauerwas, S.: 1977, *Truthfulness and Tragedy*, Notre Dame University Press, Notre Dame, Indiana.
6. Jonsen, A.: 1991, 'Casuistry as Methodology in Bioethics', *Theoretical Medicine* 12, 295–307.
7. Jonsen, A., Siegler, M., and Winslade, W.: 1982 *Clinical Ethics*, Macmillan Publishing Co., New York.
8. Jonsen, A., and Toulmin, S.: 1988, *The Abuse of Casuistry*, University of California Press, Berkeley.
9. Lindbeck, G.: 1984, *The Nature of Doctrine*, Westminster Press, Philadelphia.
10. Ogden, S.: 1986, *On Theology*, Harper and Row, San Francisco.
11. Pellegrino, E., and Thomasma, D.: 1988, *For the Patient's Good*, Oxford University Press, New York.
12. Ramsey, P.: 1970, *The Patient as Person*, Yale University Press, New Haven.
13. Siegler, M.: 1979, 'Clinical Ethics and Clinical Medicine', *Archives of Internal Medicine* 139, 914–5.
14. Tracy, D.: 1975, *Blessed Rage for Order*, Seabury Press, New York.
15. Zaner, R. M.: 1988: *Ethics and the Clinical Encounter*, Prentice-Hall, Englewood Cliffs, NJ.

SECTION I

THE MEDICAL COVENANT PAST, PRESENT,
AND FUTURE

INTERPRETING THE PHYSICIAN-PATIENT RELATIONSHIP: USES, ABUSES, AND PROMISE OF THE COVENANT MODEL

The relationship between physician and patient lies at the core of the practice of medicine. Over the centuries, that relationship has been variously interpreted, though until recently, those interpretations were largely variations on the Hippocratic tradition of medical paternalism [3, 10, 19]. In the past three or so decades, however, that approach to understanding the therapeutic encounter has undergone serious challenge. To a considerable degree, this has been due to the emergence in the United States of a concern for rights, as well as the development of a new field – bioethics. Both of these influences fostered the primacy of autonomy and its expression in the right to self-determination. Resulting from this has been a serious undermining of paternalism, a generation of alternative conceptualizations of the clinical encounter, and even some transformation of interactions between physician and patient. By interpreting the physician-patient relationship from the perspective of autonomy (and rights), the patient's status and role have assumed greater importance, whereas the physician's has been modified and, in some, cases, severely diminished.

There are increasing indications, however, that autonomy is not the last word. Further developments in the understanding of the physician-patient relationship are already underway. A re-evaluation of autonomy and the place of rights in society, the emergence of ethicists from among the ranks of clinicians, dissatisfaction with principlism (the current prevailing approach to medical ethics), and economic considerations and constraints are some of the factors contributing to the forging of new understandings.[1]

Though most interpretations of the clinical encounter have come out of the disciplines of medicine, sociology, and philosophy, theology has made its own contributions. Historically, both religion and theological reflection have exerted considerable influence upon the perception of the physician's role and responsibilities [2, 12].[2] It seems, however, that it is only quite recently that theology, at least in the United States, has attempted anything like a sustained theological interpretation of the nature of the physician-patient relationship.[3] The major work in this area has been limited to two individuals. Paul Ramsey in his *The Patient as Person* [26] and William F. May in "Code, Covenant, Contract, or Philanthropy" [16] and *The Physician's Covenant* [17] both employ the biblical/theological concept of covenant as the fundamental metaphor for understanding the clinical encounter. In "Models for Ethical Medicine in a Revolutionary Age," [33] and in much of his other work [34, 35, 36], philosopher Robert Veatch secularizes the language of covenant by employing the language of contract, claiming a near identity

3

G.P. McKenny and J.R. Sande (eds.), Theological Analyses of the Clinical Encounter, 3–27.
© 1994 *Kluwer Academic Publishers. Printed in the Netherlands.*

between the two. The language of contract in fact has virtually eclipsed the language of covenant. So while theology has addressed the nature of the physician-patient relationship, this has not been widespread, the themes it has appealed to in doing so have been limited, and the breadth, depth, and permanency of its influence are questionable.

Nevertheless, the promise of the covenant model for an understanding of the relation between physicians and patients is worth considering, particularly during this period of further development. I will attempt to do this in three steps. First, I will examine what covenant has meant and how it has functioned for the three authors previously mentioned. Second, out of this discussion, I will attempt to identify the contributions and limitations of a covenant model for understanding the physician-patient relationship. And, finally, I will briefly consider whether and how a covenant model might hold promise as a theological contribution to understanding the therapeutic relationship.

USES OF THE COVENANT MODEL

Ramsey: Covenant Fidelity and Informed Consent

Paul Ramsey is probably the first religious ethicist to interpret the therapeutic encounter in terms of the biblical and theological notion of covenant (at least in any sustained kind of way) and to probe what such an understanding might mean for specific kinds of interactions. This is not surprising, for covenant is the grounding principle of Ramsey's entire ethics. Any valid study of the nature and origin of Christian ethics, Ramsey contends, must focus on the biblical notion of God's righteousness which is rooted in the idea of covenant between God and humankind ([23], p. 2). God freely invites Israel to be God's people, requiring of them faithful obedience to God's will. God promises to be their God and maintains unwavering faithfulness to the covenant despite Israel's chronic infidelity. Such covenant-love serves as the measure for the sort of fidelity that should characterize the human response to both God and to fellow humans. With regard to the latter, Ramsey explains that "the righteousness (*tsedeq*) of God provides the measure of true justice for all human justice (*mishpat*). . . . God's righteousness becomes the plumb line for measuring the rightness of human relationships" ([23], p. 5, p. 8, p. 12). Consequently, just as God has committed God's self to human beings in faithful love, so also human beings should commit themselves to one another.

Since God's justice manifested itself in a special care for the person in need (e.g. Dt. 10:16–18; Jer. 22:3, 15–16; Ps. 72:1–4), the same should be true of human justice ([23], p. 6, pp. 12–13). Fidelity to covenant, therefore, requires an active concern for the weak and the disadvantaged, not because of anything due them, but simply because of their need. For Ramsey, "each according to his real need" sums up the biblical notion of justice ([23], p. 14). Need alone is the measure of God's righteousness toward individuals.

Commitment to the neighbor then, in the context of covenant, means the exceptionless or constant commitment to the neighbor's need. Herein lies the source of Ramsey's concern with those who are vulnerable, a theme found throughout his ethics, including his medical ethics.

The righteousness of God, while manifested in the history of the People of Israel, is most clearly manifested in the person of Jesus Christ and his preaching of the reign of God's righteousness in the Kingdom of God. For this reason, Ramsey maintains that any Christian ethics must be "decisively and entirely Christocentric." What Ramsey takes as central for his own ethics is Jesus' preaching of the Kingdom of God (particularly as expressed in the Sermon on the Mount) and the implications of that for human life and his understanding of God's love. These are the sources of Ramsey's view of the righteousness demanded of human beings, a righteousness that is most strenuous in its requirements.

Ramsey interprets Jesus' ethic as an ethic of radical love, which cannot be adequately understood apart from Jesus' expectation of the imminent realization of God's kingdom. It was because of this that he expressed the righteousness of God as the meaning and measure of human obligation in the most unqualified fashion ([23], p. 45). In the face of the inbreaking kingdom, Ramsey maintains, "preferential loves, even those justifiable in normal times, were supplanted by entirely non-preferential regard for whomever happened to be standing by" ([23], p. 39). Only the *one* neighbor whom one encounters matters and to this individual is owed disinterested love for his or her sake. This, Ramsey holds, is Jesus' teaching, and it becomes central to Ramsey's own ethic. Neighbor-love for Ramsey is the very meaning of moral obligation and, in fact, constitutes what is right. For this reason, he argues that Christian ethics is a deontological ethics ([23], pp. 115–116).

Put differently, Ramsey believes that Christian ethics is "mainly concerned about the requirements of loyalty to covenants among men, about the meaning of God's ordinances and mandates, about the estates and moral relations among men acknowledged to follow from His governing and righteous will, about steadfastness and faithfulness. . . ." ([25], p. 125). Christian life and ethics require moving deeper and deeper into the meaning of covenant obligations

to specify as aptly as possible the meaning of the faithfulness to other men required by the particular covenants or causes between us. The relevant features which this understanding of the moral law uncovers in every action, moral relation, or situation are primarily the claims and occasions of faithfulness. We are therefore driven ever deeper into the meaning of the bonds of life with life. The relevant moral features are not primarily . . . oriented . . . upon consequences. They are, rather, perceptions of claims upon us already aptly comprised in appropriate principles or canons expressive of specifiable loyalties ([25], pp. 125–126).

Ramsey does not totally discount the importance of consequences, but he emphasizes the primacy of "what makes for fidelity." It is primarily the faithfulness elements of acts, relations, and situations that are to be taken into account in determining the moral justifiability of behavior. As Ramsey puts it, "there is moral behavior whose justification is not dependent on consequences, even if not independent of consequences" ([25], p. 126). The claims of covenant faithfulness as suggested earlier are nothing other than the claims of neighbor-love. These claims, for Ramsey, give rise to principles and norms, some of which are exceptionless.

There is much more to be said about the main aspects of Ramsey's covenant ethics. These few broad strokes, however, will need to suffice as a background for considering how covenant shapes his approach to medical ethics in general, and to the physician-patient relationship in particular. Though Ramsey devoted several books and articles to medical ethics, it is in his *Patient as Person* that his covenant ethic is most explicit. Ramsey's concern in this volume is not primarily to examine the physician-patient relationship, but rather to explore a variety of "medical covenants," one of which is the physician-patient relationship. In addition, consistent with his deontological approach to ethics, he is not concerned with a description of the therapeutic encounter, but rather with identifying the right-making characteristics of these covenants and to probing several pressing *problems* in medical ethics in the light of covenant.

In the book's Preface, Ramsey explicitates one of his basic assumptions, consistent with the biblical notion of covenant: human beings are born into covenants of life with life. Being "covenanted" is, in essence, part and parcel of human existence. In addition to this more ontological covenant of all with all, people enter into specific covenants, e.g. parent-child, teacher-student, employer-employee, researcher-subject, physician-patient. The ethical task is to discover the requirements of faithfulness, of obedient love in each. The moral requirements which govern physician/researcher-patient/subject relationships, according to Ramsey, are simply a particular instance of the moral requirements that govern any human relationship, namely, the requirements of covenant fidelity ([26], p. xii) "*Justice, fairness, righteousness, faithfulness, canons of loyalty, the sanctity of life, hesed, agape or charity* are some of the names given to the moral quality of attitude and of action owed to all men by any man who steps into a covenant with another man – by any man who, so far as he is a religious man, explicitly acknowledges that we are a covenant people on a common pilgrimage" ([26], pp. xii–xiii).

Ultimately, fidelity to fellow human beings requires that they be regarded as ends in themselves and never only as means to some other end. This obligation rests, Ramsey believes, in the sacredness or inviolability of the individual in all dimensions and all stages of his or her life. Because of this sanctity, "awesome respect" is required of all human beings in their relations with one another. As a result, certain actions simply are not permitted; they are inconsistent with covenant faithfulness toward persons.

Despite Ramsey's initial emphasis on covenant and covenant faithfulness in the Preface, and his affirmation that he is speaking throughout the volume as a theological ethicist, he does, in fact, resort to somewhat secular interpretations of covenant relationships. In fact, he states that his "main appeal" is to the "community of moral discourse concerning the claims of persons"

([26], p. xii). The biblical norm of fidelity to covenant is not, he admits, a very prominent feature. Consequently, in many instances, the requirements of covenant are expressed in non-biblical language. The insights he may have arrived at through his biblical approach he communicates in terms of a natural ethic. Hence, the rules and guidelines he proffers are at times grounded in justice and at other times in agape. At play here are two levels for interpreting covenant, one natural and the other revealed.

This move is possible for Ramsey because of his view of creation. He claims, like Karl Barth, that covenant-fidelity is to be the inner meaning and purpose of our creation as human beings, and that all of creation is the external basis and condition of the possibility of covenant. "This means that the conscious acceptance of covenant responsibilities is the inner meaning of even the 'natural' or systemic relations into which we are born and of the institutional relations or roles we enter by choice, while this fabric provides the external framework for human fulfillment in explicit covenants among men" ([26], p. xii). Hence, various natural and social structures can provide the external conditions for the realization of covenant. These, for Ramsey, are worthy of preservation. While they may or may not lead to true covenant fidelity, that is their inner meaning.

The basic norm for human covenants, from the perspective of their inner meaning, is agape. But at the natural level, it is justice, though for Christians, agape should transform justice. One frequently finds in Ramsey's medical ethics a juxtaposition of the natural and revealed warrants for covenant fidelity. The latter can appeal to Christians, and the former to others.

And so one finds in *The Patient as Person* the notion of "consent" as the primary form or expression of covenant faithfulness in the physician-patient relationship. Consent is the primary requirement of loyalty to the covenantal relation between physician and patient, or researcher and subject. It is a demand of justice at the level of natural covenants, and a minimum requirement for protecting the sacredness of persons, especially the weak and the disadvantaged. Here we see evidenced the unwavering commitment to the protection of persons that lies so close to the core of Ramsey's ethics.

In response, therefore, to the question "What constitutes right action in medical practice?," Ramsey answers: "the requirement of a reasonably free and adequately informed consent" ([26], p. 2). This, for him, is the chief *canon of loyalty* between the parties in a clinical encounter. Ramsey explains:

Any human being is more than a patient or experimental subject; he is a *personal* subject – every bit as much a man as the physician-investigator. Fidelity is between man and man in these procedures. Consent expresses or establishes this relationship, and the requirement of consent sustains it. Fidelity is the bond between consenting man and consenting man in these procedures. The principle of an informed consent is the cardinal *canon of loyalty* joining men together in medical practice and investigation. In this requirement, faithfulness among men – the faithfulness that is normative for all the covenants or moral bonds of life with life – gains specification for the primary relations peculiar to medical practice ([26], p. 5).

Ramsey envisions the physician-patient relationship (and the researcher-subject relationship) as a "cooperative enterprise," a "partnership," a "joint adventuring" in a common cause. In instances of medical experimentation, the common cause is the advancement of medicine and benefit to others, whereas, in therapy or in diagnostic or therapeutic investigations, it is some benefit to the patient. Even in the latter case, there is a joint venture, according to Ramsey, in which both parties should be able to say "I cure" ([26], p. 6). Consent is necessary in both types of context because of the tendency of human beings "to overreach one side of the equation and to dominate, manipulate, or even subordinate the other for the sake of good consequences" ([26], p. 6). And it is for this reason that the consent requirement is not a once-and-for-all event, but something which is ongoing and which must be continuously repeated.

The consent requirement, Ramsey maintains, is a deontological claim and it is one without exception in ordinary medical practice (this excludes emergency situations). Although he cannot absolutely rule out some future exception in an extreme situation, he believes that the future will not reveal any morally significant exception, but will rather further disclose what this canon of loyalty means and what it entails in new situations ([26], p. 9). It simply should not be held open to any possible future exceptions.

In the grave moral matters of life and death, of maiming or curing, of the violation of persons or their bodily integrity, a physician or experimenter is more liable to make an error in moral judgment if he adopts a policy of holding himself open to the possibility that there may be significant, future permissions to ignore the principle of consent than he is if he holds this requirement of an informed consent always relevant and applicable ([26], p. 9).

It is quite likely the case, that one of the primary reasons Ramsey holds the consent requirement to be exceptionless is in order to protect the more vulnerable party in the relationship.

One of the clearest applications of covenant-fidelity in the medical context is with regard to experimentation on children and other incompetents. Ramsey contends that on the basis of consent as a canon of loyalty

children, who cannot give a mature and informed consent, or adult incompetents, should not be made the subjects of medical experimentation unless, other remedies having failed to relieve their grave illness, it is reasonable to believe that the administration of a drug as yet untested or insufficiently tested on human beings, or the performance of an untried operation, may further *the patient's own recovery*. . . . ([26], pp. 11–12).

The limits this rule imposes on practice are essentially clear: where there is no possible relation to the child's recovery, a child is not to be made a mere object in medical experimentation for the sake of good to come. The likelihood of benefits that could flow from the experiment for many other children is an equally insufficient warrant for child experimentation. . . . ([26], p. 12).

To experiment on children in ways that are not related to them as patients is already a sanitized form of barbarism; it already removes them from view and pays no attention to the faithfulness-claims which a child, simply by being a normal or a sick or dying child, places upon us and upon medical care. We should expect no morally significant exceptions to this canon of faithfulness to the child ([26], pp. 12–13).

Consent as a canon of loyalty even prohibits a competent person from consenting for the child or adult incompetent ([26], p. 14). The presumptive or implied consent of the child is not acceptable for investigations unrelated to the child's condition ([26], p. 25). "The moral issue here does not actually depend on age, but on whether anyone should be made the property of another and disposition be made of him, without his will, that is not also in his behalf medically" ([26], p. 35).

Covenant faithfulness is also expressed in the requirement of consent in the donation of organs ([26], pp. 186–197) and in the performance of transplants, particularly if they are experimental ([26], p. 219, p. 222). It takes the form of "only" caring in response to the dying (Ramsey's ethic here is not based on quality of life considerations or on self-determination), the use of a lottery in the allocation of scarce medical resources ([26], p. 265, p. 259), and, in matters of determining death, the primary patient's death alone must be kept in mind (not someone else's need for organs) ([26], pp. 108–112).

In sum, covenant is the core moral concept for Ramsey and governs all human relationships, including the medical encounter. Covenant says something about the nature of human beings – that they exist in relationships – and also about how they ought to relate – at minimum, justly and, preferably, with a faithful love aimed at meeting the neighbor's need. Agape requires a disinterested love of the other, the utmost respect for the sacredness of the other, and a special concern for and protection of the weak and disadvantaged. Covenant fidelity is the basic moral imperative in all relationships and the task of ethics is to discover what that entails in concrete situations. Any action which is not consistent with covenant fidelity cannot be right regardless of the good consequences it might produce.

Veatch: Reinterpreting Covenant into Contract

If Ramsey's understanding of the therapeutic relationship can be characterized by "covenant with a hint of contract," then Robert Veatch's can be described as "contract with a hint of covenant" (though Veatch himself might not agree with this). Two years after Ramsey published his *Patient as Person*, Veatch published an essay in *The Hastings Center Report* entitled "Models for Ethical Medicine in a Revolutionary Age" [33]. The themes he developed there reappeared in *A Theory of Medical Ethics* [34] and in *The Patient-Physician Relation: The Patient as Partner* [36].

Veatch's understanding of the physician-patient relationship can only be adequately understood in the context of his larger project, namely, the development of a universal basis for making medical ethical decisions. Professional codes, and the code of the medical profession in particular, he argues, cannot provide this ([34], pp. 82–107; [36], pp. 20–28).[4] Nor can the various sectarian theories. Instead a basis is needed that is more objective and commonly accepted than either of these two sources. How might such a universal ethic come about? Veatch contends it should be invented. To those who believe a universal basis for morality can be discovered, Veatch responds that there is no agreement that there is anything to be discovered, whether by revelation, reason, or intuition, not to mention the epistemological problems involved with all three methods. He opts instead for inventing a universal basis along the lines of a social contract ([33], pp. 110–126; [36], pp. 28–32).

What Veatch has in mind by "social contract" is not the contract of the social contractarians represented by Hobbes and Locke. Rather, he uses the term in a twofold way. One is epistemological. Social contract is a way of inventing, of coming to know, the fundamental moral principles that will govern a society. Here Veatch is influenced by Rawls' notion of social contractors coming together behind a veil of ignorance to decide upon the governing principles of their society. What emerges from this exercise, which is dependent for its success on treating everyone's interests equally and respecting each person's liberty, is what Veatch calls the "basic social contract" (the second way in which he uses the term). It consists essentially in several principles which Veatch is convinced reasonable people coming together for this purpose would choose: autonomy, fidelity, veracity, avoiding killing, and justice. The members of the moral community generated by this process are bound together, according to Veatch, by "bonds of mutual loyalty and trust. There is funda-mental equality and reciprocity, something missing in the philanthropic condescension of professional code ethics" ([34], p. 125). This is the first type of contract.

Veatch proposes two other types. These in fact are real contracts and derive from the basic social contract. The second type or level of contract is that between society and the professions, including the health professions. Any profession-specific duties or obligations cannot be derived from within the profession itself. Rather, in accordance with Veatch's theory of the basic contract, members of society, within the framework of the basic principles that have been agreed upon, might determine "that special roles such as physi-cian or patient should carry with them special rights and responsibilities" ([34], p. 131). These special duties are arrived at in the same way as the basic social contract, that is, they are those that would be agreed to by ideal, dis-interested observers who take the moral point of view. In reality, the best that can be hoped for is an approximation of the hypothetical contractual position and so the actual contract between the profession and society will, at best, approximate the ideal ([34], p. 131). Veatch proposes that the condi-tions of this second contract should be incorporated into the contract made with the professional at the time of licensure. Professionals should have a role in formulating the content of this contract, but not an exclusive role. "In the end," says Veatch, "society will outline its terms for granting the privileges of licensure" ([34], p. 133). If need be, these terms can be rene-gotiated.

The third level of contract is that between health professionals and patients. This contract as well is shaped by the moral framework derived from the basic social contract and by the role-specific duties for guiding relations between professionals and lay people. This particular contract "fills the gaps" according to Veatch. It deals with such things as the belief systems, personal values, lifestyle preferences, and views about how principles apply to specific problems that will shape the actual interaction. It is especially in this domain that the principle of liberty is operative.

Veatch opts for a contract interpretation of the physician-patient relationship, but only after considering and rejecting three other models: the engineering model, the priestly model, and the collegial model [33].[5]

In the first model, the physician functions much like an applied scientist, attempting to be free from value-considerations, and operating only in the realm of facts. He or she simply presents the facts to the patient and allows the patient to make necessary decisions. Veatch views this model as both impossible of realization (physicians cannot conform to the value-free ideal) and undesirable (it transforms the physician into a mere engineer, negates his role as a moral agent, and gets him off the hook with regard to social responsibility). In the second model (the opposite extreme), the physician operates as the "new priest," transferring his or her expertise in medical matters to expertise in moral advice. Here the physician claims excessive moral authority in the decision-making process, frequently taking decision-making away from the patient into his or her own hands. This is classic paternalism and rests upon the Hippocratic ethic. It ultimately reduces moral norms to one: Do no harm to the patient. Veatch argues that the priestly model robs the patient of moral agency and neglects other important moral norms: produce good and not harm, protect individual freedom, preserve individual dignity, truth-telling and promise-keeping, maintaining and restoring justice.

Between the extremes of the engineering and the priestly models is the collegial. Here physician and patient view themselves as colleagues "pursuing the common goal of eliminating the illness and preserving the health of the patient" ([33], p. 7). In addition to common interests, this relationship is based upon an equality of dignity and respect. Trust and confidence play a major role. Veatch considers this model to be utopian. For the most part, there is no common interest in the physician-patient relationship because of ethnic, class, economic, and value differences. Consequently, mutual loyalty and trust are false assumptions.

Veatch himself opts for the "contractual" model. He believes that a true partnership can result only when two persons from very different backgrounds find a point of mutual interest in which each can give to the other "while retaining substantial autonomy" ([36], p. 4). Autonomy is key to understanding Veatch's insistence upon contract as the appropriate model for interpreting the therapeutic encounter. Unlike earlier times when patients were seen as "helpless, infantile, passive recipients of the professional's largess," they are now increasingly regarded as "autonomous agents capable of active participation in partnerships for the pursuit of their rights as well as their well-being" ([36], p. 47). Another reason for the choice of this model is Veatch's belief that medical decisions are essentially evaluative and that the appropriate values to guide them are not those inherent in medicine, but rather those held by the individual patient. If this is true, "then an entirely new understanding of the patient-physician relation is in order – one in which the basic ethical norms are articulated by an agreement among the members of the moral community and applied at a second level in an agreement between the various professions and the lay public. The result is a contractual understanding of medical ethics and a resulting relationship between lay person and professional that must be viewed as a true partnership" ([34], p. 279).

And what, for Veatch, does contract entail? Several points need to be made here. First, Veatch virtually identifies contract and covenant. The language of covenant for him is simply religious language for the language of contract ([34], p. 121). Covenant is a type of contractual relationship, except that special emphasis is placed "on the morally binding communal qualities of mutual loyalty and reciprocity that are essential to all contractual relationships" ([34], pp. 125–126).[6]

Second, Veatch reads the Jewish and Christian traditions as being fundamentally contractarian. "The people of Israel were created by a contract (or covenant, to use the more traditional term)" ([34], p. 121).

Third, Veatch has continuously asserted that when he employs the term contract, he does not have in mind a "legalistic or businesslike relation between physician and patient" ([36], p. 38, p. 14; [33], p. 7; [35], pp. 109–110). In fact, if there is to be a contractual ethics for medicine, these implications must be avoided.

Fourth, what Veatch does have in mind is a more "symbolic usage" of the term "as in the traditional religious or marriage 'contract' or 'covenant'" ([36], p. 14). In using this comparison, he seems to want to maintain both the interpersonal bond *and* some degree of social regulation for the protection of the parties.

Marriage is a contract that is fundamentally a relationship of trust and loyalty, of bonding between two people. We surely could ask no more of the patient-physician relationship. This is the kind of contract I have in mind.

But we are aware that the marriage contract, spiritualized and romanticized as it is, is embedded in a social, political, and even legal structure. If things fall apart, as unfortunately they do from time to time, there is a socially solid hard core set of institutional structures to fall back upon, not just a mystical, ethereal bond. That is to me what covenant has always required; it is contract at its best" ([35], p. 110).

In specifically describing the patient-physician relationship as contract, Veatch writes:

Here two individuals or groups are interacting in such a way that there are obligations and expected benefits for both parties. These obligations and benefits are limited in scope, though, even if they are expressed in somewhat vague terms. The basic principles of autonomy, fidelity, veracity, avoiding killing, and justice are essential to a contractual relationship. The premise is trust and confidence, even though it is recognized that there is not a full mutuality of interests. Social sanctions institutionalize and stand behind the relationship, in case there is a violation of the contract, but for the most part the assumption is that there will be a faithful fulfillment of the obligations ([36], p. 14; [33], p. 7).

A contractual approach to the physician-patient relationship takes account of the many divergences between the parties and builds in mechanisms to control for them. "It emphasizes mutual obligations to inform, the right of either professional or lay person to withdraw when tensions emerge, the necessity of placing the relationship in a larger social context of licensing boards, state laws, etc." ([35], p. 111).

Fifth, the language which Veatch employs to describe contract in many ways echoes the language generally used to characterize covenant. He speaks, for example, of the premise of this contractual relationship being "trust and confidence" and of the basic assumption being "faithful fulfillment of the obligations." It must emphasize the "moral equality between the partners," and "fidelity to promises made" ([36], p. 38). The norms for the relationship "are trust, loyalty, respect, and faithfulness, not legalism and business bargaining" ([36], p. 38). If trust and confidence are broken, the contract is broken ([33], p. 7).

Sixth, near the core of Veatch's understanding of the contract between patient and physician is autonomy, the moral agency, of both parties, but especially of the patient. The contract is largely about shared decision-making with each party maintaining their integrity. This requires the mutual establishment of the basic value framework for making decisions. Only with this mode of relationship, Veatch believes, can there can be genuine sharing of ethical authority and responsibility. In his most recent book on this subject, *The Patient-Physician Relation: The Patient as Partner*, Veatch puts it this way:

... The patient-physician relation ought to be one in which both parties are active moral agents articulating their expectations of the interaction, their moral frameworks, and their moral commitments. The result should be a partnership grounded in a complex contractual relation of mutual promising and commitment ([36], p. 3).

This kind of partnership based on shared ethical authority and responsibility is not true of either the engineering or the priestly models, according to Veatch. And the collegial model is problematic because it gives the false impression of an equality between patient and physician.

Seventh, Veatch's view of the physician-patient relationship, though stressing autonomy, is not wholly individualistic. He claims, in contrast, that "the partnership about which we have been speaking can never truly be limited to the individual level of two isolated actors pursuing mutually acceptable goals. It must always be embedded in a social ethic" ([36], p. 159). Veatch's social ethic is deontological, as is his individual ethic. While the latter gives precedence to autonomy over patient benefit, the former gives priority to justice rather than to the good of the greatest number. This means for Veatch striving for the common good or maximizing societal benefits "within the constraints of the uniqueness of individuals as equals in their claim on social resources" ([36], p. 161).

Eighth, and finally, the norms which govern the relationship do not come from within the profession nor even from within the relationship itself. They are largely derived from the basic social contract. In addition, these norms, as mentioned above, are deontological norms. Veatch insists that there is an absolute lexical ordering of non-consequentialist principles over beneficence. Consequences do not count morally until principles have been satisfied ([36], p. 6).

May: Covenant Indebtedness

Like Veatch, May finds the Hippocratic ethic to be an inadequate basis for considering the nature of the physician-patient relationship and the duties and obligations that result therefrom. The reasons for his discontent, however, are different from Veatch's. So is his project.

May's interest is in probing the various images that shape the physician's view of the world in which he or she functions and which defines his or her professional role, in view of identifying the one which best captures the meaning of the physician-patient relationship. The choice of images in May's view is not insignificant, for images and metaphors have both a disclosive and a formative effect. They both reveal and shape reality. While May is interested in both aspects, his greater interest seems to lie with the latter. "The task of ethics in the professional setting," he writes, "might be called, at least in part, corrective vision" ([17], p. 13). It attempts a "knowledge-able revisioning of the world" as a corrective to a warped vision. A distorted vision of reality and of our duties within it is likely to lead to problematic behavior as well as to a further shaping of the world in the image of the warped vision ([17], pp. 14–15).

In exploring the images that disclose the physician-patient relationship, May sets out to speak theologically and to take seriously the contribution of religious discourse to medical ethics. Most philosophers and theologians working in medical ethics, in May's judgment, work either as secular moral-ists or as closet Christians and Jews. Their religious convictions are suppressed as resources because they perceive both society and the profession of medicine to be "confessionally indeterminate" ([17], p. 25). May strongly disagrees with this approach.

While culturally understandable, this response, in my judgment, mistakenly diagnoses our times. For better or for worse, the modern world reeks of religion. These religious forces do not always, or even for the most part, conform to official Judaism and Christianity, but they pervade our times and, not least, those fateful events which attend sickness, suffering, and death. These events shatter or suspend the ordinary resources that people trust for managing their lives and send them to the doctor in hope of rescue. They clothe the doctor accordingly in the images of shelter and rescue – the parent, the fighter, and others. The full power of these images and the hold that they have over the layperson and therefore the professional does not become clear except that we see them, at least in part, in a religious setting ([17], p. 25).

In May's view, the current religious setting is a negative one. It is defined by the central anxiety of modern man: anxiety before death, the fear of catastrophic death, a fear of bodily destruction and of abandonment by the community. "In effect, preoccupation with death and destructive power has replaced attentiveness before a good and nurturant God as the central religious experience of modern people" ([17], p. 32). In response to this shift, the formerly dominant image of the physician as parent has been replaced by the image of the physician as fighter. From this perspective, the physician's authority derives from the fear of suffering and death and from the powers against it which modern biomedicine puts at his or her disposal.

In light of this, May's objective is to suggest an alternative vision, a corrective vision, one grounded in a basic biblical image, namely, covenant.[7] He writes in *The Physician's Covenant*:

This book . . . will . . . explore the alternative religious vision that the West derives from the biblical tradition. That tradition affirms the holy of holies to be creative, nurturant, and donative, rather than destructive. It does not deny the reality of disease, suffering, and death but puts them in the context of a power that transcends them. God ultimately encompasses disease and death. An excellent term to describe the peculiar "tie" or "bind" of men and women to the sacred so perceived is the biblical term "covenant." The covenant illuminates the patterns of resistance, avoidance, and worship of death and reveals that, while they differ in many ways, they still make the same metaphysical mistake: they all take death too seriously; they assume that death has the last word, that death defines, without significant remainder, the healer's task. This noncovental religious reflex ultimately falsifies the goals of medicine and compromises the ties between the patient and the healer ([17], pp. 34–35).

May's interest, therefore, lies not only in an interpretation of the physician-patient relationship, but also in the deeper more fundamental metaphysical context in which physicians and patients interact. That context for May is profoundly theological. Suffering, illness, and death invariably bring one to the threshold of the religious. Attending to this dimension can supply a corrective vision both for the experience of illness and for understanding the doctor-patient relationship. How does May work this out?

In his initial foray into the domain of the physician-patient relationship, May deals with three models: code, contract and covenant [16]. His work is essentially a critique of the Hippocratic approach to interpreting the physician-patient relationship that has dominated modern medicine. May observes that the Oath includes three parts. The first consists in codal duties toward patients – prohibitions and positive obligations that are essentially philanthropic in their origin. The second specifies the covenantal obligations that physicians owe their teacher and their teacher's family. These consist in accepting full filial responsibility for the teacher's (the physician's adoptive father) personal and financial welfare, and transmitting without charge his knowledge and his art to his teacher's and his own progeny, as well as to other students. The third part of the oath is the setting of the first two parts within the context of an oath "by the gods."

May agrees with the historian Ludwig Edelstein in his characterization of the oath: the duties which physicians have toward their patients are expressed as an "ethical code," whereas, the duties assumed toward the members of one's profession are viewed as a "covenant." The reason for the difference, May believes, lies in the fact of indebtedness. Physicians are indebted to their teachers for what they have received from them, hence, use of the term covenant to describe that relationship. The dynamic here harkens back to one of the chief characteristics of the biblical notion of covenant – gift/response. May puts it this way: "Physicians undertake duties to their patients, but they *owe something* to their teachers. They have received goods and services for which they owe their filial services. Toward their patients, they function as benefactors, but toward their teachers, they relate as beneficiaries. This responsiveness to gift characterizes a covenant" ([17], p. 110; [16], p. 31). Just as in biblical covenant the duties of the people are placed within the context of the gift of divine deliverance, a gift already received, so likewise the physician undertakes obligations to his teacher and his teacher's progeny out of gratitude for services already rendered. May maintains that modern medicine has reinforced this ancient distinction between code and covenant and that it has in fact opted for code as the norm for the physician-patient relationship.

This distinction, May contends, is not good for the moral health of the profession. The language of codes has a different connotation than the language of covenant. The differences truly suggest a different kind of relationship. Furthermore, a code approach to the therapeutic relationship says nothing about indebtedness. Covenantal indebtedness does not enter into the conception of physician duties toward patients. May also finds the third part of the oath problematic. "The religious oath partly resembled a covenant in that the physician made a promise that referred to the gods from whose power the profession of healing ultimately derives. To this degree it put the physician in the position of a recipient" ([17], p. 111). This vow, however, does differ from biblical covenant in that there is no mention of prior divine actions which elicit the human response, and it diminishes the responsive element because the physician swears *by* the gods rather than promising *to* the gods.

What disturbs May even more than the codal definition of the doctor-patient relationship is its philanthropic basis. It "assumes that the professional's commitment to patients is a wholly gratuitous rather than a responsive act" ([17], p. 112). The codal approach to the physician's duties toward patients recognizes neither an indebtedness to a transcendent source nor any indebtedness of the physician to the community. "As a result," says May, "an odor of condescension taints the documents" ([17], p. 112). May refers to this as the "conceit of philanthropy." It allows for no reciprocal sense of indebtedness between the profession and particular patients or the community. The professional's commitment to patients is a purely gratuitous act rather than one that is responsive or reciprocal. "The code offers the picture of a relatively self-sufficient monad, who out of the nobility and generosity of his disposition and the gratuitously accepted conscience of his profession, has taken upon himself the noble life of service" ([17], p. 113).

The medical codes ground the duties of patients to physicians on what they have received from physicians. A section of the code emphasizes the many benefits which society has received from the profession which constitute the reason for its indebtedness to the profession. But there is nothing in the codes which ground duties of the profession to society on the basis of indebtedness for benefits received. There is even, as May notes, a shift in language in the 1847 AMA code. It refers to the *duties* of physicians toward their patients, but *obligations* of patients toward their physician. The latter suggests an *indebtedness* to the profession for services rendered, while the former suggests an acceptance of duties to patients out of a noble graciousness. It does not view itself in any way as beholden.

May's claim is that physicians are, in fact, beholden. They are indebted to their patients and to the community in a variety of significant ways. A covenantal ethic assists in the acknowledgement of this reality and in fostering a sense of mutual indebtedness. As an image it both helps to disclose and to shape. As disclosure, it draws attention to indebtedness. And for May there is no doubt that physicians owe a considerable debt to the community for, among other things, society's investment in medical training, making available teaching hospitals, and patients offering themselves for experimentation or for teaching. The physician, claims May, is beneficiary as well as benefactor. "Only within a fundamental responsiveness do professionals undertake their secondary little initiatives on behalf of others" ([17], p. 116).

In emphasizing the elements of exchange and reciprocity, does May in effect suppress the element of gratuity, thereby transforming covenant into contract? "Does covenant," Mays asks, "simply act as a commercial contract in which two parties calculate their respective best interests and agree upon some joint project from which both derive roughly equivalent benefits for good contributed by each?" ([17], pp. 116–117).

May's response to this is important for while opting for covenant he does recognize the appeal and the value of contract which he sees as threefold. First, it establishes some symmetry between physician and patient as they work toward an explicit or assumed agreement about the exchange of goods. Also, it fosters respect for human dignity and autonomy, insists on informed consent, and encourages the specification of rights and duties as delimitations of the contract. Second, contract allows for legal enforcement of the terms of the contract, thereby offering protection to both parties and serving as an incentive to make each accountable. Third, in relying on self-interest, contract does not depend on philanthropy or on a condescending charity. Each party enters the contract because each has something to gain.

Despite these advantages, however, May believes it would be unfortunate if professional ethics were reduced to a commercial contract "without significant remainder." Why? There are several reasons. First, a contract approach to the therapeutic relationship tends to "reduce professional obligation to self-interested minimalism, *quid pro quo*" ([17], p. 118). The gift dimension in human relationships is lost or severely diminished. May believes that a contract cannot take into account the unexpected in the clinical encounter which may call for more than what has been specified in the contract. It is impossible to delimit in advance what needs may develop. Second, covenants, unlike contracts, are internal, cutting deep into personal identity. They are ongoing, enveloping, and affect a person's identity. Someone initiated into the profession of healing is always a healer. He or she is not a healer only when performing certain tasks. When viewed in this way, the profession of healing entails an element of giving and receiving that develops and increases. This is not true of a contract. In sum, "the kind of minimalism that a purely contractualist understanding of the professional relationship encourages produces a professional too grudging, too calculating, too lacking in spontaneity, too quickly exhausted to go the second mile with patients along the road of their distress" ([17], p. 122). Third, contract medicine fosters not only a minimalism but also a certain kind of maximalism. That is, there is a tendency to offer treatments to the patient which are really not necessary, but are useful only for protecting the physician. As with minimalism, the motive here is self-interest. Fourth, because of the emphasis on self-interest, a contractualist ethic must rely on external constraints to keep physicians within moral limits. Because marketplace controls do not work in the context of illness, and, consequently, the patient is always at a disadvantage, May wants to see internal fiduciary checks. Fifth, contracts determine only what is required, and not what is just. This can give the advantage to the more powerful party in the relationship, the physician. But the biblical notion of covenant requires that the more powerful be responsible to the needs of the less powerful. "It does not permit free reign to self-interest, subject only to the capacity of the weaker partner to protect self through knowledge, shrewdness, and purchasing power" ([17], p. 124). Contractualism offers few restraints on action other than those explicit in the contract and prudent self-interest.

Despite Veatch's distinguishing between a commercial and a primordial social contract, May nevertheless considers his approach to be inadequate. Veatch's understanding of the human is not sufficiently communal, and without a greater sense of the communal nature of human beings and of the common good, there would seem to be little to constrain individuals from compromising the principles arrived at through the original position in view of self-interest ([17], pp. 125–26). The ideal of rational self-interest is unlikely to hold up in the real world with its destructive forces. Furthermore, Veatch treats religion as quite dispensable. He fails to take account of its metaphysical dimension and what that suggests for an understanding of the world. How might such an understanding make a difference? May writes:

The healer nurtured in the Christian understanding of covenant affirms the holy of holies as creative, nurturant, and donative rather than destructive. This affirmation does not deny the reality of disease, pain, suffering, and death but puts them in the setting of a power that persists and endures in the very midst of them. God ultimately encompasses suffering and death; they are *real* but not *ultimate*; they do not speak the last word about the human condition. The Christian sees in Jesus an event that does not eliminate suffering and death . . . but instead exposes destructive power in its final impotence to separate men and women from God ([17], p. 127).

This covenantal context frees healers from the need to avoid ties with those who are perishing. Relationship with the transcendent makes other relationships possible.

Detached from this setting, however, the ideals of technical proficiency, philanthropy, and contract tend to deteriorate into devices whereby healers, beset by pain, pettiness, and suffering, shield themselves from patients and their perishing life. . . . Contractors thus dart in and out of the patient's world of need, shoring up their own life through the transaction of selling. Contractors guard their own interests, specifying carefully the precise amount of time and service for sale. Thus code, philanthropy, and contract, within the context of death, are all devices for evading ties. All have in common a fear of perishing, of drowning in the plight of the other. Ties suck one down into the vortex of death. To call contract, or any other of these devices, merely a secular version of covenant overlooks the important question of metaphysical setting" ([17], pp. 128–129).

Furthermore, the transcendent grounding of a covenant ethic resolves a potential conflict between responsiveness and gratuitousness. Unlike the wholly gratuitous altruism of philanthropy, covenantal ethics is responsive. And unlike the emphasis on self-interest in contractualism, it is gratuitous. Response to debt and gratuitous service, while seeming contradictory, are not when viewed in the context of the transcendent. This is because gratuitous service is itself response to what has been received from the transcendent.

Finally, it needs to be said that May does not totally reject either code or contract. He views covenant rather as encompassing these two and going beyond them. Codes foster the transmission of technical skills and this is important. Contracts allow for the enforcement of obligations and even legal redress when physicians are remiss in carrying out their duties to patients. Covenant fidelity includes both of these dimensions and more.

THE PROMISE OF THE COVENANT MODEL

In light of this review of the use of covenant by Ramsey, Veatch, and May in their interpretations of the physician-patient relationship, what can be said about the promise of this model as a theological contribution toward illuminating the therapeutic relationship? That is, what are its contributions and limitations as exemplified in the work of these three thinkers? Attempting to answer this question is somewhat perilous for it presupposes some normative concept of covenant as well as some normative understanding of the clinical encounter. Neither of these exists. Richard Zaner's *Ethics and the Clinical Encounter* [38] is perhaps one of the most thorough and illuminating phenomenologies of the physician-patient relationship, but it can hardly be said to be normative. The difficulty is compounded when one looks for a normative understanding of the covenant model. What are its essential features? There is, unfortunately, no pure example of covenant, no agreed upon set of characteristics that can be attributed to the model. Even in the biblical materials, there is a plurality of covenants ranging from the promissory covenants made with Noah (Gen. 9:8–17), Abraham (Gen. 15, 17), and David (II Sam. 7), to the "law covenant" of Sinai (Ex. 19:1–20; 24:3–8) and Horeb (Dt. 4:44–6:3), to the implicit references to covenant in the prophets, to the New Covenant personified in Jesus Christ. Adding to the difficulty is the controversy among biblical scholars regarding the origin and nature of many of these alleged covenants [14]. Joseph Allen in his *Love and Conflict: A Covenantal Model of Christian Ethics* [1], suggests three common features of all biblical covenants: the parties to the covenants are unequal in status; God alone initiates and establishes the terms of the covenants; a response to God's initiative is required. Even if these common features were agreed upon, are they sufficient, and what would be their implications for understanding covenant relationships?[8] Finally, there is the danger of attributing to the model itself what is actually a feature of the model's use by a particular thinker. Despite these difficulties some attempt will be made to probe the contributions and limitations of the covenant model for interpreting the physician-patient relationship.

The first issue to be addressed is the version of covenant being considered. Is it a secular version or a biblically grounded version? While Ramsey's and May's understandings of covenant are rooted in the biblical witness, Veatch's in the end is not. His transformation of covenant into contract is more than merely linguistic. It is not simply the substitution of a secular term for a religious one. The notion of covenant is different. Though Veatch does speak of loyalty, trust, promising, and faithfulness as characterizing contracts/covenants, his concept lacks a transcendent grounding and referent. Also absent are the gift/response character of biblical covenant, the sense that human covenants are grounded in a prior covenant of God with humankind, the normativity of the divine-human covenant for human relationships, and the concern for the disadvantaged and the needy. When all is said and done, contract/covenant for Veatch seems to mean the interaction of autonomous moral agents for the pursuit of mutual self-interest, guided by the principles of the basic contract and the duties derived from it. While this may be a plausible interpretation of the physician-patient encounter and may incorporate some features of biblical covenant, it is not truly a biblical/theological understanding of covenant and should not be considered as such.

At least one of the ways in which Veatch's understanding of contract/covenant is important, however, especially for those employing a covenant model, is the way in which it highlights the place which law can and probably should play in the doctor-patient relationship. Trust, mutuality, faithfulness and other characteristics of covenant are undoubtedly primary, but also crucial is the specification of responsibilities and obligations and, in some instances, their legal enforcement. Such a legal dimension to the physician-patient relationship is not inconsistent with the biblical reality which, in fact, contains a legal dimension. And it may well be a necessity given the modern context of the physician-patient relationship – one in which the parties to the encounter are frequently strangers, litigation is a constant threat, the profit motive sometimes rules, and the like. In order to ensure the presence of this aspect of the relationship, however, it is not necessary to resort to contract.

Another consideration in assessing the promise of the covenant model is the degree to which it "fits" or accurately and adequately describes the encounter between doctors and patients. Several interpreters of the therapeutic relationship note that the very core of that relationship is need and response – the need of the patient for healing and the obligation of the physician, because he or she is physician, to seek the patient's good "even at some cost to the comfort, power, prestige, or fiscal benefit of the physician" ([28], p. 31). As Pellegrino and Thomasma describe it "medicine as a human activity is . . . a response to the need and plea of a sick person for help, without which the patient might die or suffer unnecessary pain or disability. . . . It is grounded in the claim that comes from the vulnerability and suffering of a fellow human" ([28], p. 32). Richard Zaner [38] offers a similar analysis of the physician-patient relationship. He describes several features of the experience of illness which in large part constitutes the moral foundation of medicine. To experience illness is generally to be someone who needs assistance and who appeals for help. Such an appeal usually involves asking health professionals to put their knowledge, time, energies, and experience in the service of healing. This encounter, according to Zaner, necessarily evokes and requires trust on the part of the patient who places him or herself in the hands of others, and responsive care from health professionals. Zaner also observes that the caregivers "in whom the patient invests unavoidable trust and reliance are almost invariably strangers" (p. 54) and that the encounter is almost always marked by asymmetry of power in favor of the health professional because of the vulnerable state of the patient ([38], p. 55).

What the covenant model captures and preserves is precisely those primary features of the physician-patient encounter described by Pellegrino, Thomasma, and Zaner – the inequality of the relationship, the vulnerability of one of the partners, the dynamic of need and response, the necessity of trust, the pursuit of the vulnerable party's good as the primary imperative, and a certain open-endedness in the response depending on the individual's original and changing needs. Both Ramsey's and May's use of covenant in one way or another captures these elements. Hence, it would seem that covenant rather than contract, at least Veatch's understanding of contract, seems better suited to express what at least some and probably many believe to be core ingredients and dynamics of the doctor-patient encounter. For this reason, the covenant model serves a crucial function. It focuses attention on critical dimensions of the relationship and thereby can assist in shaping the self-understanding of both physician and patient as well as the interaction between the two. It can serve as a "corrective vision," to use May's term, against the many influences that can and do distort the relationship, e.g. an excessive focus on autonomy and the resulting individualism.

There is yet another important way in which the covenant model can serve as "corrective vision." This contribution, as well, is illustrated in both Ramsey's and May's use of it. Biblical covenant is a relational reality which encompasses a variety of relationships that run vertically as well as horizontally. In addition, it includes a profound sense of the infinite worth and sacredness of human beings, based on religious reasons. The person in his or her dignity is always in relation to the Other, to individual others, and to the community of others. Implied in this relatedness are responsibilities to the multiple others with whom the individual is in relationship. There is in biblical covenant no autonomous individual nor is autonomy the central feature of any relationship. The emphasis is on right relationships, on just relationships, rather than on the exercise of autonomy over against another with few ties to the community of others.

There are echoes of this biblical perspective in both Ramsey and May, though in May it is more implicit. For Ramsey, covenant discloses the relational nature of human beings and covenant fidelity the necessity of conducting those relationships in such a way that they take account of the sanctity of personal life. There is value in Ramsey's deontological emphasis: human beings, because of their divinely grounded sanctity, should not be used, manipulated, or subordinated to other ends. It is essential that physicians recognize the sacredness that is the patient before them, the patient who is in need of special protection because of his or her vulnerability in this relationship. Covenant fidelity focuses the physician's attention on the patient primarily and does not allow the patient to be subordinated to other interests, be they economic, professional, or social. And by underscoring the infinite value of all and the equality of all, it provides a bulwark against any social assessment of the worth of individuals and their lives. Ramsey's intense concern for protecting the individual in the medical context, however, especially the vulnerable individual, contributes to a rather individualistic perspective, inconsistent with the covenant concept. This individualism is further reinforced by Ramsey's particular interpretation of neighbor love. It should be noted that Ramsey's inclination toward individualism occurs for mostly theological reasons and not because of an emphasis on autonomy and/or individual rights. Informed consent does not devolve from autonomy, but rather from a spelling out of covenant fidelity in a certain type of context and from Ramsey's theological beliefs about persons.

Because May's purposes and language are different from Ramsey's, his concern for protecting the sacredness that is the person, in particular the vulnerable patient, and for the individual's ties to the larger community are somewhat less evident. It can undoubtedly be gleaned from the way in which he speaks of individuals, especially patients, and in his critique of a contractualist ethic as inherently not attending sufficiently to "the more vulnerable and powerless of the two parties" ([17], p. 124). His community-oriented perspective can be seen in his attempt to argue the indebtedness of physicians both to their patients and to the community, and in his critique of Veatch's contractarianism. "Its conception," he says, "is insufficiently communal. . . . Does not the original self require a more communal sense of humankind, a more spacious sense of the common good . . . ?" ([17], p. 125). It is further evidenced in May's discussion of the distribution of health services and of covenanted institutions, neither of which we have taken up in this essay ([17], pp. 137–141, 169–176).

These aspects of the covenant model constitute extremely important contributions to an understanding and shaping of the physician-patient relationship. In this sense, the covenant model does much. But there is also a sense in which it achieves too little. This is not an indictment of the covenant model itself, but rather a recognition that no one model is entirely adequate to illuminating, understanding, and shaping the encounter between doctor and patient. In what ways does covenant achieve too little?

While covenant concentrates on the nature of the interaction between physician and patient, it does virtually nothing to illuminate the contexts of the interaction, particularly the clinical context – its multiple participants, its components, its variability, and its dynamic. Physician-patient interactions do not take place in the abstract; they are situated. And it is precisely the context which provides the texture of the relationship. To speak of covenant fidelity or covenant indebtedness is ultimately not enough. This insufficiency is particularly a problem for Ramsey and Veatch. Both proceed deontologically and so deductively. Ramsey regards the physician/researcher-patient/ subject relationship as just another instance of the more general covenant governing relations among people. His goal is ultimately to formulate principles that express the requirements of covenant fidelity that apply without exception. Veatch proposes a universal basis for ethics, consisting in a set of principles that can then be applied to specific situations. In both cases, the deductive approach results in a neglect of the clinical context. While the covenant model is not inherently deontological, there is really nothing about it which would require attention to context. May's early work, for example, is not deontological, yet does not deal with the multiple aspects of the clinical encounter other than the character of the interaction between physician and patient. Where there is little or no attention given to the clinical setting, covenant appears to be something imported from without and imposed upon the doctor-patient relationship, thereby placing its appeal in some jeopardy.

A further limitation of the covenant model is that it may tend to be either reductionistic or amorphous. Ramsey reflects the former and May, to some degree, the latter. For Ramsey, as we have seen, the core requirement of the encounter between physician and patient, researcher and subject is "reasonably informed consent." While important, is this what is most important in the clinical encounter and is it enough? Are there no other moral requirements in the relationship? It seems that speaking only of consent as the primary requirement of covenant fidelity impoverishes both covenant fidelity and the interaction. Much the same can be said of Veatch's notion of contract/covenant. It is reduced to a set of principles of which the most important is autonomy. May, on the other hand, tends to be a bit too vague about what covenant entails in the physician-patient encounter. It does mean giving and receiving, going the extra mile, not imposing upon patients what is unnecessary, developing internal fiduciary checks, pursuing what is just, contributing to the common good. But what does each of these mean and entail and are there other crucial characteristics of the clinical encounter in addition to these?

Finally, because covenant implies an inequality in relationships and requires particular attention to be given to the less powerful in the relationship, the covenant model tilts the interpretation of the physician-patient relationship in favor of the patient. In both Ramsey and May (and even in Veatch), there is a sense of protecting or empowering the patient over against the physician. This may have been necessary in view of the paternalism of the past and may continue to be necessary in light of the vulnerability of most patients, but it would be both unfortunate and misguided to neglect the relationship (and the responsibilities) of the patient to the physician, to the health care community, and to society. The implications of covenant in this regard have not been sufficiently explored, though it is not entirely clear how much the covenant concept can contribute.

Covenant is a rich concept, particularly in its theolological version, and it can offer much toward understanding core dimensions of the interaction between physicians and patients. This is especially true at a time when patient autonomy has widely been taken as the defining characteristic of the relationship, and when legal and economic factors are increasingly coming to shape it as well. The covenant model can serve as "corrective vision." It can broaden understandings of the relationship, while centering attention on some of its central and indispensable features. It interprets the clinical relationship with a breadth and depth that the contract model does not. And because the biblical concept of covenant carries with it a certain worldview, it is capable of offering a way of providing meaning to some basic human experiences.

But if the biblical concept of covenant is going to be an effective metaphor for understanding and shaping the clinical encounter, it must also encompass the insights of other appropriate models and metaphors, and pay more serious attention to the realities of the clinical setting.[9] Covenant is not sufficient in itself, nor can it remain abstract and disembodied. It must, for example, rigorously probe what it means to entrust one's well-being to a stranger and what faithfulness means between a doctor and a patient in the current climate of health care delivery. Even if it were successful in these respects, there would still remain a nagging question: Who would listen? But that is another problem for another time.

The Park Ridge Center
and
The Department of Religion, Health, and Human Values
Rush University
Chicago, Illinois, U.S.A.

NOTES

[1] While it is probably true that too much can be made of models of the physician-patient relationship and that the actual experience can get lost in debates about which is most appropriate [7], nonetheless, models are not unimportant. Particular interpretations not only describe varying aspects of the interaction between physician and patient and underscore its essential characteristics, they can also serve to shape how the interaction is perceived and what occurs within it. A contractual model *is* different from a covenantal model. Each has its own features. Each emphasizes certain qualities of character and modes of behavior. Both in turn differ from a priestly or paternalistic model and from other proposed models. How one conceives the nature of the relationship does make a difference in how one conceives the goals of the encounter, the roles of each of the participants, their respective duties and responsibilities, and the dynamics of the interaction.

[2] As David Kelly [12] points out in his study of the development of Catholic medical ethics in the U.S., the manuals for confessors in the fifteenth and sixteenth centuries, as well as the later manuals of moral theology, contained a special section on the moral obligations of physicians.

[3] Theology no longer occupies the position of influence in medical ethics that it once held. As the field began to develop in the mid-1960s in response to a multitude of developments in the life sciences, theologians were among the first to engage the issues, issues which themselves were seen as being religious in nature. The prominence of theology was short-lived, however. Within a relatively brief period of time, philosophy and law emerged as the dominant players and conversations partners, to the virtual exclusion of theology. One of the primary factors contributing to this development was the growing need for the formulation of public policy with respect to many of the new biomedical technologies. Religious language, and arguments based on religious convictions, have been considered inappropriate and even a hindrance in developing policy in a pluralistic society. Philosophy and law together provided the language and a mode of moral reasoning based on rationality that would be more objective, more impartial, and, hence, have more universal appeal. The result of these efforts, as Daniel Callahan observes, has been "a mode of public discourse that emphasizes secular themes: universal rights, individual self-direction, procedural justice, and a systematic denial of either a common good or a transcendent individual good" ([6], p. 2, [11]. This, by and large, is the context in which discussions of a covenantal interpretation of the physician-patient relationship have taken place. And while the winds of change are blowing through U.S. bioethics and are affecting its conception, its methodology, and its practice, this continues to be part of the present context in which the question is raised: What can theology, and in particular the concept of covenant, contribute to an understanding of the clinical encounter?

[4] Veatch rejects the consensus of health care professionals as a basis for an ethic, as well as the belief that only those within a profession can know what is required in the professional role.

[5] One gains some insight into Veatch's personal inclination toward the contract model from a comment he makes in his essay "The Case for Contract in Medical Ethics": "Trained in the tradition of liberal political philosophy with its social contract and in the theological tradition of covenant, which I recognized as providing the historical roots for secular contract, I adopted the term *contract* as a counter to the romantic convergence theory of the collegial model, a term borrowed from Talcott Parsons" ([35], p. 109).

[6] Commenting on his own use of contract language in contrast to May's use of covenant, Veatch writes: "There is almost nothing in May's essay to which I can take any exception. If there is any difference at all it would be in my concern that May . . . and all others who are nervous about contract language . . . overly spiritualize and romanticize the relationship making it as spineless as the buddy-buddy-relationship of the collegial model. I have consistently made clear that I oppose individualistic and legalistic explications of the contract metaphor. But I am also concerned about the squishy, apolitical, asocial romanticism of the collegial model" ([35], p. 110). Veatch here seems to misunderstand the covenant concept. It is not apolitical or asocial,

and it need not be squishy. There is in the biblical concept a strong sense of specific obligations arising out of relationships and expressed in laws.

[7] Early on in his discussion of covenant, May identifies several structural features of biblical covenant: an original gift between the soon-to-be covenanted partners (deliverance from Egypt); a promise based on the gift (the vows at Sinai); and the acceptance by the people of a set of ritual and moral obligations. "These commands," says May, "are both specific enough to make the future duties of Israel concrete (e.g., the dietary laws, and laws governing protection of the weak), yet summary enough (e.g., 'Love the Lord thy God with all they [sic] heart. . . .') to require a fidelity to intent as well as to letter" ([17], p. 109). This last statement is particularly important in view of Veatch's concern that covenant is "squishy." Law and the specification of obligations is a part of biblical covenant and May recognizes this.

[8] Actually, Allen identifies three common features of covenant relationships, whether the covenant relationship into which all are born or specific covenant relationships. They require respect for the dignity of persons, an obligation to be concerned for the needs of covenant partners, and faithfulness ([1], p. 56).

[9] In some ways, the work of Zaner [38] and Pellegrino and Thomasma [28] exemplify what is being suggested here, though their accounts are surely more philosophical than theological.

The Park Ridge Center for the Study of Health, Faith and Ethics
Chicago, Illinois
U.S.A.

BIBLIOGRAPHY

1. Allen, J.: 1984, *Love & Conflict: A Covenantal Model of Christian Ethics*, Abingdon Press, Nashville.
2. Amundsen, D., and Ferngren, G.: 1983, 'Evolution of the Patient-Physician Relationship: Antiquity Through the Renaissance', in E. E. Shelp (ed.), *The Clinical Encounter*, D. Reidel Publishing Co., Boston, pp. 1–46.
3. Bloom, S.: 1978, 'Therapeutic Relationship: Sociohistorical Perspectives', in W. Reich (ed.), *Encyclopedia of Bioethics*, vol. 4, The Free Press, New York, pp. 1663–1668.
4. Bouma, H., et. al.: 1989, *Christian Faith, Health, and Medical Practice*, Eerdmans, Grand Rapids.
5. Cahill, L.: 1975, 'Paul Ramsey: Covenant Fidelity in Medical Practice', *Journal of Religion* 55 (October), 470–476.
6. Callahan, D.: 1990, 'Religion and the Secularization of Bioethics', in D. Callahan, and C. Campbell (eds.), 'Theology, Religious Traditions, and Bioethics: A Special Supplement', *Hastings Center Report* (July–August), 2–4.
7. Carson, R. A.: 1988, 'Paul Ramsey, Principled Protestant Casuist: A Retrospective', *Medical Humanities Review* 2, no. 1 (January), 24–35.
8. Clouser, K. D.: 1983, 'Veatch, May, and Models: A Critical Review and a New View', in E. E. Shelp (ed.), *The Clinical Encounter*, D. Reidel Publishing Company, Boston, pp. 89–103.
9. Emanuel, E. J., and Emanuel, L. L.: 1992, 'Four Models of the Physician-Patient Relationship', *Journal of the American Medical Association*, 267, 2221–2226.
10. Entralgo, P.: 1978, 'Therapeutic Relationship: History of the Relationship', in W. Reich (ed.), *Encyclopedia of Bioethics*, vol. 4, The Free Press, New York, pp. 1655–1663.
11. Fox, R.: 1989, 'The Sociology of Bioethics', in *The Sociology of Medicine*, Prentice-Hall, Englewood Cliffs, NJ, pp. 224–276.

12. Kelly, D.: 1979, *The Emergence of Roman Catholic Medical Ethics in North America*, The Edwin Mellon Press, New York.
13. McCarthy, D. J.: 1963, *Treaty and Covenant: A Study in Form in the Ancient Oriental Documents and in the Old Testament*, Pontifical Biblical Institute, Rome.
14. McCarthy, D. J.: 1972, *Old Testament Covenant: A Survey of Current Opinions*, John Knox Press, Richmond.
15. Masters, R. D.: 1975, 'Is Contract an Adequate Basis for Medical Care?', *Hastings Center Report* 5, no. 6 (December), 24–28.
16. May, W. F.: 1975, 'Code, Covenant, Contract, or Philanthropy?', *Hastings Center Report* 5, no. 6 (December), 29–38.
17. May, W. F.: 1983, *The Physician's Covenant*, Westminster Press, Philadelphia.
18. May, W. F.: 1991, *The Patient's Ordeal*, Indiana University Press, Bloomington.
19. Mechanic, D.: 1978, 'Therapeutic Relationship: Contemporary Sociological Analysis', in W. Reich (ed.), *Encyclopedia of Bioethics*, vol. 4, The Free Press, New York, pp. 1668–1672.
20. Meilaender, G.: 1989, 'On William F. May: Corrected Vision for Medical Ethics', *Second Opinion* 10, 104–124.
21. Mendenhall, G. E.: 1970, 'Covenant Forms in Israelite Tradition', in E. Campbell and D. N. Freedman (eds.), *The Biblical Archaeologist Reader*, vol. 3, Doubleday, Garden City, New York, pp. 25–53.
22. Quill, T. E.: 1983, 'Partnerships in Patient Care: A Contractual Approach', *Annals of Internal Medicine* 98, 228–234.
23. Ramsey, P.: 1950, *Basic Christian Ethics*, University of Chicago Press, Chicago.
24. Ramsey, P.: 1967, *Deeds and Rules in Christian Ethics*, Charles Scribner's Sons, New York.
25. Ramsey, P.: 1968, 'The Case of the Curious Exception', in G. Outka and P. Ramsey (eds.), *Norm and Context in Christian Ethics*, Charles Scribner's Sons, New York, pp. 67–135.
26. Ramsey, P.: 1970: *The Patient as Person*. Yale University Press, New Haven.
27. Pellegrino, E., and Thomasma, D.: 1981, *A Philosophical Basis of Medical Practice*, Oxford University Press, New York.
28. Pellegrino, E., and Thomasma, D.: 1988, *For the Patient's Good: The Restoration of Beneficence in Health Care*, Oxford University Press.
29. Smith, D. H.: 1985, 'Medical Loyalty: Dimensions and Problems of a Rich Idea', in E. E. Shelp (ed.), *Theology and Bioethics*, D. Reidel Publishing Co., Boston, pp. 267–282.
30. Smith, D. H.: 1987, 'On Paul Ramsey: A Covenant-Centered Ethic for Medicine', *Second Opinion* 6 (November), 106–127.
31. Sturm, D.: 1985, 'Contextuality and Covenant: The Pertinence of Social Theory and Theology to Bioethics', in E. E. Shelp (ed.), *Theology and Bioethics*, D. Reidel Publishing Co., Boston, pp. 135–161.
32. Thomasma, D.: 1992, 'Models of the Doctor-Patient Relationship and the Ethics Committee: Part One', *Cambridge Quarterly of Healthcare Ethics* 1, 11–31.
33. Veatch, R.: 1972, 'Models for Ethical Medicine in a Revolutionary Age', *Hastings Center Report* 2, no. 3 (June), 5–7.
34. Veatch, R.: 1981, *A Theory of Medical Ethics*, Basic Books, Inc., New York.
35. Veatch, R.: 1983, 'The Case for Contract in Medical Ethics', in E. E. Shelp (ed.), *The Clinical Encounter*, D. Reidel Publishing Company, Boston, pp. 105–116.
36. Veatch, R.: 1991, *The Patient-Physician Relation: The Patient as Partner, Part 2*, Indiana University Press, Bloomington.
37. Werpehowski, W.: 1991, 'Christian Love and Covenant Faithfulness', *The Journal of Religious Ethics* 19, no. 2, 103–132.
38. Zaner, R. M.: 1988, *Ethics and the Clinical Encounter*, Prentice-Hall, Englewood Cliffs, NJ.

WILLIAM F. MAY

THE MEDICAL COVENANT: AN ETHICS OF OBLIGATION OR VIRTUE?[1]

A covenantal ethic, above all else, defines the moral life responsively. Moral action (such as selling, refraining, respecting, or giving) ultimately derives from and responds to a primordial receiving.

In its ancient and most influential form – the biblical covenant – covenantal obligation arises from specific exchanges between partners that lead to a fundamental promise which, in turn, defines the identity and therefore shapes the future of both parties to the agreement. The biblical covenant included the following four elements:
1. an original exchange of gifts, labor, or services;
2. a promise based on this original or anticipated gift;
3. the acceptance of an inclusive set of moral obligations that will shape the future life of both partners; and
4. the provision of ritual and other means for the renewal of life between partners in the course of their subsequent alienation from one another.

The scriptures of ancient Israel are littered with such covenants between men and women and between nations, but they are controlled and judged throughout by that *singular covenant* which embraces all others: the covenant between God and Israel. The latter includes the aforementioned elements: first, a gift (the deliverance of the people from Egypt); second, an exchange of promises (at Mt. Sinai); and third, the shaping of all subsequent life in response to the original gift and the promissory event. God "marks the forehead" of the Jews forever, as they accept an inclusive set of moral commandments by which they will live. These commands are both specific enough to make the future duties of Israel concrete (e.g., laws governing protection of the weak) yet summary enough (e.g., "Love the lord thy God with all thy heart. . . .") to require a fidelity that exceeds a legalistic specification. Fourth and finally, the biblical narrative marks out those ritual means whereby Israel returns regularly to those foundational events that shape her life (the dietary laws, the Sabbath and the holidays). These elements variously reappear in the horizontal covenants between sovereigns and subjects, treaties between nations, and the all important covenant of marriage. For Christians, God's covenant with Israel structurally prefigures the inclusive covenant that will spread across the whole of humankind in God's Son [3, 4, 5].

The subsequent meaning of the word has not always carried forward the biblical sense of a covenant. The word, indeed, has often referred to a variety of unsavory practices – real estate covenants that keep blacks or Jews out of particular neighborhoods; loyalty to the professional guild which sometimes

29

G.P. McKenny and J.R. Sande (eds.), Theological Analyses of the Clinical Encounter, 29–44.
© 1994 *Kluwer Academic Publishers. Printed in the Netherlands.*

takes precedence over professional duties to patients and clients; country club agreements that build invisible walls around the playing fields of the well-favored; hiring practises and referral systems that toss jobs and business to those with an "in." Further, the religiously disposed have often confused the divine covenant with an utterly parochial loyalty, reducing God to their gender, race, nation, class, neighborhood, or kinship group. The ancient prophets condemned the resulting amalgam and confusion as the worst sort of idolatry, the worship not of the idol that frankly tempts at a perceptible moral distance from the holy of holies, but of the counterfeit that slyly mounts the altar in the sanctuary itself and wraps itself in a stolen glory.

In its biblical form, the concept of a covenant offers resources for criticizing the narrowness and exclusivity of these various types of idolatry. God, the creative, nurturant, and donative source of all beings, establishes the primary covenant that measures all others. Loyalty to God, whatever its particular implications, requires loyalty to all of God's creatures. Thus the covenant that distinguishes Jews and Christians from others requires them at the same time to deal open handedly with others – not only with familiars but also with strangers. The prophets, Amos and Isaiah, and the gospel writer, Luke, the physician, make abundantly clear that the biblical covenant must open outwardly toward others in servant love. The primary covenant with God serves as a critical standard for all those lesser covenants that people enter into and that tempt them to turn away from the stranger and the needy.

In this essay, I reflect on the bearing of this four-part, covenantal, story line on the medical covenant and, even more particularly, to reflect on whether covenantal thinking leads to an emphasis in moral theory on obligations or virtues. That particular assignment invites me to jump one way or another into the current debates on the subject of theoretical ethics. The field of ethics conventionally requires coverage of at least three topics: the moral agent, the agent's action and the results of that action. Virtue theorists usually concentrate on the agent and his or her character; obligation theorists focus either on the agent's action (duty theorists) and/or on the results of that action (utilitarians or pragmatists).

While I will explore the implication of a covenantal ethic for the virtues, I am not prepared to claim that such an ethic can do without obligation theorists such as Paul Ramsey and James F. Childress who have developed principles and rules to shape medical practise. Covenantal fidelity defines the ruling principle in Ramsey's pioneering work, *The Patient As Person*, which he has rigorously applied to such issues as experimentation on human subjects, high risk therapy, organ transplants, care for the dying, genetic engineering, abortion, and in vitro fertilization. Childress, also a religious obligation theorist, has co-authored *The Principles of Biomedical Ethics*, the most influential book in the field, that appeals to the principles of non-maleficence, beneficence, respect for autonomy, and justice as the comprehensive ground for all subsidiary moral rules in biomedical research and practise.

The influence of these obligation theorists derives largely from their excellence, but from other factors as well, religious, philosophical, societal and organizational. First, religiously, one can hardly dismiss the efforts of obligation theorists to mine the biblical covenant for principles and rules. The tradition, after all, recognizes that the power that binds us (*re-ligio*) elaborates that binding in the form of moral obligations (*ob-ligatio*) stated in the form of divine commands which would shape human action and produce results. Ramsey believed that faith in a promise-keeping God readily converts into a general canon of loyalty and the derivative principle of the sacredness of life. Childress believes that the ancient divine commands that prohibit, for example, killing, stealing, lying, and injustice or that enjoin good works readily organize under the general principles of non-maleficence, respect for autonomy, justice, and beneficence, those principals which he believes should shape medical practice.

However, neither author in his influential work spends much time on the biblical covenant. (Ramsey deals with the source expeditiously in his preface; and Childress observes, but only in passing, that some deontologists derive moral duties directly from the commands of God, while others appeal to the dictates of reason, common sense, or intuition.) Both seem to be convinced philosophically and theologically that the basic moral commands are relatively detachable from the commander. This uncoupling of narrative and moral principle serves the Christian ethicist who wants to throw light on vexing moral issues in biomedicine, but within the general framework of philosophical discussion. Indeed, Childress collaborated with the utilitarian, Tom L. Beauchamp, in writing *The Principles of Biomedical Ethics*, on the ground that ethicists who differ in their theoretical justifications can nevertheless agree on the principles. (This approach awakens memories of the Anglican, Richard Hooker, who distinguished between what is necessary and accessory to salvation; Childress tends to relegate underlying beliefs to the accessory in the moral life.) Different derivations of principles do not appear to make much difference at the level of content.

Obligation theorists also believe that their approach allows the religious ethicist to address the society at large and the professions in particular without divisive appeals to different foundational commitments. Christian ethicists, through mediating principles, can throw light on vexing moral issues in medicine, while not losing a larger professional and societal audience, only some members of which share Christian belief. Ramsey and Childress' books have powerfully influenced the practise of medicine, a profession which generally thinks of itself as confessionally indeterminate.

The funding sources for research in biomedical ethics have also been largely confessionally indeterminate. Not the churches and synagogues, but huge secular foundations and the federal government have largely bankrolled research in medicine. Money does not necessarily bark and command, but it does sweet talk and beckon, as it defines both a field and the acceptable linguistic currency in that field.

Finally, a society dominated by large scale organizations finds itself particularly in need of generally acceptable moral principles. (Sheldon Wolin, the political scientist, covered the last 200 years in his final chapter of a single volume history of Western political theory under the rubric, "The Age of the Organization" [7]. In a similar vein, Alasdair MacIntyre once observed that Max Weber, the student of bureaucracy, not Karl Marx or Adam Smith, ranks as the ruling theoretician of our time.) Huge institutions that deliver services to large numbers of people, often times strangers, need general policies that will help produce reliable behavior upon which not only specialized, and therefore mutually dependent, colleagues but also their patients can rely. Understandably, then, ethicists who articulate and elaborate standardizing principles and rules answer a felt need both for practitioners and policy-makers in large institutions and for the afflicted who seek conscientious treatment there. One tends to expect these standards to be relatively acceptable and independent of the accidents of personal biography, philosophical conviction, and religious disposition.

For these several reasons, I cannot dismiss the agenda of obligation theorists for medical ethics. Principle-oriented theory has some validity, philosophically and religiously; and it serves variously a religiously pluralist and secular society dominated by large organizations. But I also believe that a covenantal ethic must attend to the question of the moral agent and his or her virtues. Covenanted men and women do not simply accept a set of rules and principles guiding their actions, they also bind themselves over in the course of an event which alters and continuingly defines their identity. At Mt. Sinai, God marked the forehead of the Jews forever; through baptism, Christians acquire their very name and identity. As a covenanted people take up the particulars of their several vocations, their identity will display itself in the virtues that ought to characterize their practise. No ethic that adequately explores the medical covenant can focus simply upon the quandaries that emerge in medical practise and the bearing of moral principles upon those quandaries; it must also explore the identity of those agents who profess medicine. Virtue theory is the name we give to ethics as it focuses on such questions of identity and character.

Principle-oriented moralists concentrate on the question, "what should we do?" Virtue theorists pose a second question that lies behind the first, "whom shall we be?" Practitioners' answer to the latter question of identity may more fatefully affect, for good or for ill, their actual practise than the articulation of those principles which obligation theorists associate with applied ethics. What does it mean to be a physician? Is the physician simply a hybrid, an aggressive mix of technician and entrepreneur, or something more, a professional? If the practitioner is simply a careerist who puts his skills up for grabs to the highest bidder, then answers to the questions of truth-telling, price-charging, guild-policing, and resource allocation will simply reflect at best the driving force of enlightened self-interest. The practitioner simply sells time and services, throwing in an occasional act of charity to air blow the image and salve the conscience. Only in acknowledging his or her identity as a professional does the practitioner assume the full burden of those moral principles which obligation theorists emphasize.

PROFESSIONAL IDENTITY

"To profess" means "to testify on behalf of," "to stand for," or "to avow" something that defines one's fundamental commitment. A profession opens out toward an as yet unspecified transcendent good that defines the professor. In contrast, a career is self-referential; a career reminds us of the word "car," a self driven vehicle. We all breathe easier if we have a career that supplies us with a kind of self-driven vehicle through life whereby we enter into the public thoroughfares but for our own private reasons and with our own private destination in mind. The Hippocratic Oath acknowledged a transcendent element in the profession of healing when the young physician avowed, "I swear by Apollo Physician, and Asclepius and Hygieia, and Panaceia and all the gods and goddesses, making them my witnesses, that I will fulfill according to my ability and judgment this oath and covenant" ([2], p. 6). In the spirit of the Hippocratic tradition, legend has it that a particularly dedicated professor of history at a university used to answer the phone, not with his name, but with, "History here." The activity had so taken hold of the man as to burn away the self-referential. The story would have been less charming had he answered, "Historian here." The controlling activity rather than personal attainment, is the point. The conventions of our telephone transactions don't permit it, but what the stricken patient hopes for in calling the doctor is, "Healing here."

What does it take to make good on that answer? Abraham Flexner, the mother of all reformers in medical education, identified six distinguishing marks of a professional which, in my judgment, reduce to three: intellectual, moral, and organizational. Each of these marks, in turn, calls for a primary virtue. The three marks and their correlative virtues set the agenda for the rest of this essay.

THE INTELLECTUAL MARK AND PRUDENCE

A professional draws on a complex and esoteric body of knowledge not available to everyone but to which he or she has direct access. This direct access to first principles distinguishes professional education from mere training. Trained people can perform specific routines but, argued Flexner, they don't know *why* they perform them. They quickly lapse into undeviating patterns; they don't adapt or grow. A knowledge of first principles, however, lays the foundation for the professional's future growth.

The need for direct acquaintance with basic principles led Flexner to insist on locating professional education in universities and on closing the so-called proprietary schools which smacked of mere training. He further insisted on the university site for medical education since a profession must, in some range of its work, engage in scientific research which expands the profession's knowledge base and thus serves its advancement and progress. Not any and all professionals must be researchers, but somewhere within the profession – usually at the university – the research enterprise must be supported.

If the physician's intellectual mark consists of scientific knowledge, no more, then the student might need little more than the virtue of perseverance to acquire it. While a lowly virtue, perseverance is indispensable to the acquisition of scientific knowledge under the trying conditions of lengthy professional education today. But healing is an art, not simply a science, requiring something more than the virtue of perseverance. Additionally, the physician needs the virtue of prudence.

Contemporary medical literature, however, offers little help on the meaning of the phrase, "the healer's art" that would help us connect the art with prudence. When practitioners plump for medicine as an art rather than a science, they sound rather apologetic, as though they want to defend a place for themselves and their slender store of experience in turf largely occupied by scientists and the huge warehouses of knowledge and equipment over which they preside in tertiary care centers. Alternatively (and sometimes patronizingly), scientists accept healing as an art only in the provisional sense that they cannot yet reduce any and all cases to investigative techniques. Had we fuller knowledge, then the instinctive surmise would disappear. Only because of gaps in our scientific knowledge do we need to venture into the intuitive and the imaginative. Thus the art of healing provides but a temporary station on the way to a more perfect science of healing.

Healing, as an art, like other arts, displays a cognitive and a creative aspect; it is both a knowing and a doing. The cognitive component in the healing process – both diagnostic and prognostic – acts partly as science but partly as art. As science, it requires specialized and abstract investigatory work. Each medical specialty rests on its peculiar knowledge base; it relies on a series of indicators that signal the presence of diseases that its technical interventions can affect. This specialized knowledge and activity systematically abstracts from the patient as a whole, by ignoring the technically irrelevant.

But the cognitive component of healing includes a further element. The healer must attempt not only to cure the disease but address the illness of which the disease forms a part. As Dr. Eric Cassell has put it, "Disease is something an organ has; illness is something a man has" ([1], p. 48). The host may be incidental to the disease (sometimes, not always), but the host is rarely incidental to the state of illness. The healer who would "make whole" stricken patients cannot rest content with a specialized and abstract knowledge, as much as that may contribute to the enterprise. The healer must look at the whole patient, the full range of somatic and psychic structures disrupted by disease. Such a knowledge must ultimately unite and specify, situating the disease in a particular person and in his or her idiosyncratic social history.

The knowledge resembles more the coordinations of an experienced fisherman scanning the sea or a hunter wary in the woods or a cook fully experienced in the kitchen than it resembles the reductions of a particular case to a general law.

Finally, the healer's knowing aims at doing. The healer's art creates as well as knows, and the constructive activity in which the healer engages includes more than curing disease. The fully rounded work of healing reconnects the patient with the world and recovers his or her self-control and self-confidence. The treatment plan at its best offers a coherent total program for as much recovery as a particular patient can achieve under the circumstances. Preventive medicine and chronic care, just as much as acute and rehabilitative medicine, require artistic intervention. Admittedly, such care has its routines, techniques, and tricks. No art form lacks its conventions. But these activities ultimately aim to reorder comprehensively a human life. That work, in the nature of the case, must unite and specify.

The fully developed physician needs the virtue of prudence since the physician's act of knowing is artful as well as scientific. Ethical theorists who orient to principles sometimes tend to downgrade the importance of prudence as they rely on general principles rather than the concrete insight of the moral agent. Some theorists trivialize prudence into a merely adroit selection of means in the pursuit of ends, into a crafty packaging of policies. The virtue of prudence, to be sure, deals with fitting means to ends. But, as a virtue, it consists of much more than the "tactical cunning" to which Machiavelli and many in the modern world diminish the virtue. Thomas Aquinas noted,

> . . . in order that a choice be good, two things are required. First, that the intention be directed to a due end . . . Secondly, that man take rightly those things which have reference to the end: and this he cannot do unless his reason counsel, judge, and command aright, which is the function of prudence and the virtues annexed to it. . . . (*Summa Theologica*, Pt. 1–11, Q. 58, Art. 5, Trans. Dominican Fathers).

The ancients gave a primary place to prudence as the "eye of the soul." The medievalists ranked prudence as the first of the cardinal virtues on the grounds that "Being precedes Truth and . . . Truth precedes Goodness" ([6] p. 4). Diagnosis and prognosis precedes apt therapy. One must discern what is there in order to be there for it. An openness to being underlies both being good and producing the good in and for others. The art of prudent discerning includes three elements, especially if one hopes to discern fully a human being:

a) *memoria* – being truly open to the past (rather than retouching, coloring, or falsifying the past);

b) *docilitas* – defined, not as a bovine docility, but as an openness to the present, the ability to be still, to be silent, to listen and take in what makes itself present; and

c) *solertia* – a readiness for the unexpected, the novel, an openness to the future, a disposition sometimes in short supply in those who only too quickly subsume the new case under old routines.

This disciplined openness to the past, present, and future fairly summarizes what the distressed patient needs from the healer and the moral tradition that argued for it long precedes Freudian wisdom on the subject of the therapeutic relationship.

Such prudence demands much more than a facile packaging of what one has to say or do. *Discretio* presupposes metaphysical perception, a sense for what the Stoics called the fitting, a sensibility that goes deeper than tact, a feel for the other and one's own behavior that is congruent with reality. Without discernment, the professional does not deal with the whole truth in diagnosis and prognosis, that is, with the truths of illness, as well as disease; and without discernment, the professional does not mobilize the full power of healing, which exceeds that of laser, knife, and drug.

THE MORAL MARK AND FIDELITY

Professionals who stand for, or avow something, do so on behalf of someone. They profess not simply any and all knowledge, but a particular body of learning that applies tonically to a specific range of human problems. Unlike earlier knowledge merchants, such as wizards and magicians, who may use their knowledge simply to display their virtuosity, professionals must orient altruistically to serve human need, and not simply their own needs or those of their friends and relatives, but those of the stranger. "Hanging out one's shingle" symbolized from the seventeenth century forward the professional's readiness to direct knowledge to the needs of the stranger.

The symbol of the shingle is ambiguous. It invites the stranger, but it does so, at least partly, in the setting of the marketplace. The professional differs from the amateur in that the amateur does it for love and the professional for money. Only a species of angelism would argue otherwise. Professionals are not disembodied spirits. They earn their bread and pay their bills in the course of serving others. The professional exchange with patients and with the institutions for which they work partly conforms to a marketplace exchange of buying and selling. It therefore requires the traditional marketplace virtues of industry, honesty, and integrity.

But the professional exchange also transcends or ought to transcend, the case nexus. It requires the further virtue of fidelity. A sustaining commitment to the being and well-being of the patient presumably distinguishes it from the marketplace assumptions about an exchange between two relatively knowledgeable, frankly self-interested parties. Ultimately, the interest and well-being of the patient must trump the physician's own self-interest. The patient's own perceptions of self-interest will not adequately protect the patient. Usually, patients cannot obey the marketplace warning, "buyer beware," because an asymmetry exists between the professional's knowledge and power and the patient's relative ignorance, powerlessness, and oftentimes urgent distress which does not permit comparative shopping. This imbalance requires that the professional exchange take place in a fiduciary setting of trust that transcends the marketplace assumptions about two wary bargainers. Only the physician's fidelity to the patient in the disposition of his or her knowledge and power justifies that trust.

Fidelity to the patient should constrain the physician's notion of what he or she has to sell. If I walk into a Volvo showroom, I do not expect the salesperson to question whether I actually need a Volvo. No one in a Volvo showroom has ever suggested to me that, as an academic, I ought to trot across the street and buy at half the price a Toyota Tercel. The salesperson takes it as a challenge to sell me a car and thereby meet his quota for the month. I am a pork chop for the eating. But if I visit a surgeon, I must be able to assume that he or she sells two items, not one. The surgeon is not simply in the business of selling hernia jobs, but also the further, detached, disinterested, unclouded judgment that I need that wretched little procedure. Otherwise, the physician abuses disproportionate knowledge and power and poisons a fiduciary relationship with distrust. Instead of sheltering, the surgeon takes advantage of the distressed.

Herein lies the ground for professional strictures against conflicts of interest. The professional must be sufficiently distanced from his or her own interests and that of other patients to serve the patient's well being. However, serving the patient's welfare faces subtler pitfalls than gross conflicts of financial interest. Physicians must take care that their eagerness to recruit their patients into high risk research protocols does not obscure their primary duty to their patients' welfare. Otherwise, they act as double agents, pretending concern for their patients while serving, in fact, the drug companies or the advancement of their own careers as researchers.

The professional exchange differs in a second way from a marketplace transaction. In addition to its disinterestedness, it is, for want of a better word, transformational, and not merely transactional. The healer must respond not simply to the patient's self-perceived wants but to his or her deeper needs. The patient suffering from insomnia often wants simply the quick fix of a pill. But if the physician goes after the root of the problem, then she may have to help the patient transform the habits that led to the symptom of sleeplessness. The physician is slothful if she dutifully offers acute care but neglects preventive medicine.

The term "transformation," however, awakens legitimate fears of paternalism. The prospect of the physician, the lawyer, the manager, and the political leader engaged in transforming habits provokes memories of overbearing authority figures who haunt the American past.

Transformational leadership slips into parentalism unless teaching becomes its chief instrument. Teaching becomes one of the few ways in which one can engage in transformation while respecting the patient's intelligence and power of self-determination. No one can engage properly in preventative, rehabilitative, chronic, and even terminal care without teaching his or her patients and their families.

This fidelity to the patient in his or her deeper needs brings the discussion back to the virtue of prudence. Prudence is sensitive not only to ends but to fitting means to those ends. The end of medicine is healing, but the art of healing must pass through rational self-determining creatures. In securing patient compliance, in acute care and patient partnership in the tasks of preventive, rehabilitative, and chronic care, the physician must honor fully the person who hosts the disease. The physician does not succeed in these tasks unless she lets herself be taught by the patient in the course of the interview and teaches the patient aright in turn. Teaching aright poses the question: how does one teach the sick and the stricken? If teaching is the means, what are the means to the means? How does one teach? Pedantically? Naggingly? Vindictively? Scoldingly? Scripture answers that question concisely: speak the truth in love. Therewith prudence and fidelity fully knit together.

THE ORGANIZATIONAL MARK AND PUBLIC-SPIRITEDNESS

This mark traces back to the medieval guild and, still further back, to the Hippocratic covenant of the young physician with his teacher who helped initiate him into his professional identity. Continuing with this tradition, Abraham Flexner believed that physicians should organize in order to maintain and improve professional standards. A professional guild ought to differ from a trade association in that it aims at self-improvement, not self-promotion. So goes the ideal.

Until recently in the United States, it appeared that the organizational mark had all but faded, except for purely defensive and self-promotional activities upon the part of the guilds. In addition to the general American myth of the self-made man, the medical profession subscribed to the myth of the free-lance physician who ran his own office, made the rounds of his patients at their homes, and managed a practise partly by cash and partly by charity. But since the 1930's medicine has increasingly organized to produce medical goods in the setting of large hospitals and to expect compensation for the distribution of these goods and services through huge third-party payers. (The professional guild has been much less disposed to accept responsibility for controlling the quality of medical goods through the mechanism of self-regulation and discipline.)

This organizational mark calls for the professional virtue of "public-spiritedness," which I would define as the art of acting in concert with others for the common good, in the production, distribution, and quality control of health care.

Why do professionals owe something to the common good? Physicians, along with other professionals, are rulers in a society such as ours. We do not transmit power today on the basis of blood. We largely wield and transmit a knowledge-based power acquired in the university. This derivation of power from a university inescapably generates an indebtedness to the society. A

huge company of people contribute to the making of practitioners as they
zigzag their way through college and professional schools: the janitors who
clean the johns, the help in the kitchen, the secretaries who make the operation hum, the administrators who wrestle with the institution's problems, the
faculty who share with students what they know, the vast research traditions
of each of the disciplines that set the table for that sharing, and the patients
who lay their bodies and sometimes their souls on the line, letting young
physicians and surgeons practice on them in the course of perfecting their
art. And behind all this, the public moneys and gifts that support the
enterprise, so much so that tuition money usually pays for only a small fraction
of the education. When professionals treat education as providing them with
a private stockpile of knowledge to be sold on the market to the highest bidder,
they systematically distort and obscure the social origins of knowledge, and
therefore the power which that knowledge places within their grasp.

Further, the power which modern professionals wield vastly exceeds that
of their predecessors. What physicians do fatefully affects the society at large.
They have even less reason than their predecessors to bend their power to serve
purely private, entrepreneurial goals. When they do so, they conform to
Aristotle's definition of tyranny, that is, the channelling of public power to
private purposes. Somehow we have managed to normalize power directed
to private purposes alone under the conditions of a marketplace democracy
without recognizing the despotic element in its exercise. It ill behooves
professionals to sever their calling from all questions of the common good
and to instrumentalize it to private goals alone. The social source of power's
derivation and the public consequences of its exercise should elicit from
professionals the virtue of public-spiritedness, the art of acting in concert
with others for the common good.

Public-Spiritedness and the Production of Medical Goods

In professing a body of knowledge which they place at the service of human
need, physicians do not do so as purely private performers. Medicine today
is increasingly a social art practised by a health care team in the setting of a
very large institution. Until recently, the hospital has served as the physician's workshop, where he or she performed services and made money under
a third party payment system with few controls over either those services or
the money to be made. However, hospitals now operate under the restriction
of a federal payment system, which controls the amount of money that the
hospital (and indirectly the doctor) can make off a given disease, and under
growing pressures from insurance companies and from corporations negotiating health care contracts. Physicians must increasingly accept some
responsibility for not only their personal professional values but also the values
shaping the institution in which they work. In effect, physicians need to
recognize the hospital as a *polis*, a political entity, in which they must help
set policies on the relative weight of ends served and the resources deployed

to reach those ends. The ill require for their healing not only covenanted individuals but covenanted institutions that profess, testify on behalf of, or stand for something. Such institutions, in turn, need physicians skilled not simply in the art of medicine, but in the art of acting in concert with others for the common good.

Physicians need the virtue of public-spiritedness for the further reason that, whether in the hospital, the clinic, or in group practice, they rarely work today as solitary gunslingers, but as members of a health care team. Medical education and residency training programs have only partly recognized the importance of cultivating the professional as a member and leader of the health care team. They concentrate exhaustively and almost exclusively on educating students in the sciences and developing their technical skills, while paying much less attention to their maturing as team members.

Further, the very conditions of residency training often distract and disrupt effective teamwork. The constantly changing caste of specialists, attending physicians, residents and nurses often leaves patients and their families wondering just who is their physician. The system also can signal to the impressionable resident that team membership is incidental and subordinate to technical performance. The ever shifting composition of the team with rotation shifts makes it difficult to deliver continuous, coherent service to a particular patient, however conscientious the team members may be. The accidental intersection of a variety of needed technical services does not spontaneously mesh into a coherent program of care. Clearly, the effective team requires not only self-conscious efforts to prepare young members for participation and leadership but also the rethinking of schedules and agendas so as to honor the team's unitary responsibilities.

Public-spiritedness and the Distribution of Medical Goods

The notion of the just and public-spirited professional calls for more than a minimalist commitment to what the moral tradition has called *commutative* justice, that is, the fulfillment of duties between two private parties based on contracts. Public-spiritedness suggests a more spacious obligation to *distributive* justice which seeks to meet the health care needs of all members of society, a goal which has not been, and cannot be, met solely through the mechanism of the marketplace and the supplements of charity. Physicians and other professionals have a duty to distribute goods and services targeted on basic needs without limiting those services simply to those who have the capacity to pay for them. Some would deny this obligation to distributive justice altogether. Others would argue that the services of the helping professions should be distributed to meet such basic human needs, but that this obligation rests on the society at large and not on the profession itself. This approach, which argues for a tax-supported third party payment system, would eliminate *pro bono publico* work as a professional obligation.

Still others, myself included, perceive the obligation to distribute professional services to be both a public and a professional responsibility. Professionals exercise power through public authority. The power they wield and the goods they control are of a public magnitude and scale. Although the state has a primary responsibility for *ministering* justice (an old term for distributive justice), professional groups, too, have a ministry, if you will, to perform. When the state alone accepts responsibility for distributive justice, a general sense of obligation tends to diminish in the culture at large, and the social virtue upon which the impetus to distribute depends loses the grounds for its renewal. Professionals particularly need to accept some responsibility for ministering justice to sustain that moral sensibility in the society at large. Otherwise, the idealistic motives that originally prompted the founding of institutions (such as mental hospitals) will peter out, and the institutions deteriorate into custodial bins.

Further, such *pro bono publico* work does not merely serve the private happiness of those individuals who receive services; it eventually redounds to the *common* good and fosters public happiness. Those who receive help are not merely individuals but *parts* of a whole. Thus the whole, in so serving its parts, serves its own public flourishing; it rescues its citizens mired in their private distress for a more public life. In the absence of *pro bono publico* work, we signal, in effect, that only those people who can pay their way into the marketplace have a public identity. To this degree, our public life shrinks; it reduces to those with the money to enter it. Public-spirited professionals not only relieve private distress, they help preserve our common life, in a monetary culture, from a constant source of its own perishing.

Public-spiritedness and Professional Self-regulation and Discipline

Physicians need the virtue of public-spiritedness not only to perform satisfactorily on health care teams in hospitals and to distribute the good of health care widely, but also to control the quality of the good produced and distributed. Professional self-regulation and discipline is the name for that quality control.

For many reasons, physicians duck responsibility for professional self-monitoring. Like any professional group, they find themselves in complex, interlocking relations with fellow professionals which make actions against a colleague awkward, the defense of abused patients inconvenient. Unlike lapses on the part of other professionals, the physician's therapeutic misadventure can be fatal or irreparable in its effects. Since the stakes are so high in the case of a physician's malpractice, professionals are tempted to draw their wagons into a circle to protect a challenged member. Unlike lapses on the part of other professionals, the physician's misadventure can irreparably harm or kill. Further, Americans, in general, press charges against their neighbors or colleagues only reluctantly; they are morally averse to officiousness. Unlike some of their European counterparts, Americans show little stomach for playing amateur policeman, prosecutor, judge, when they themselves are not directly or officially involved in an incident. In many respects, this *laissez faire* attitude is an admirable trait in the American character.

Yet this morally attractive nonchalance cannot justify permissiveness in professional life, for professionals are power wielders, *de facto* rulers in modern society. They reserve the right to pass judgment (in professional matters) on colleagues or would be colleagues; and the society supports this right in establishing educational requirements and licensing procedures. To be sure, patients profit from this through higher standards, but the profession also profits – handsomely – in money and power. Professionals have not justified this state created monopoly if they merely practice competently themselves. The individual's license to practice depends upon the prior license to license which the state has to all intents and purposes bestowed upon the guild. If the license to practice carries with it the obligation to practice well, then the prior license to license carries with it the obligation to judge and monitor well. Not only individuals but a guild must be accountable. Am I my colleague's keeper? The brief answer is, yes. And that responsibility may sometimes require not simply disciplining the troubled colleague, but finding positive ways to help him. Public-spiritedness in this always disturbing activity of professional discipline calls for the attendant virtues of courage, fairness, and compassion.

In my opening comments on the biblical covenant, I mentioned the elements of gift, promise, and moral and ritual action. The discussion of prudence and fidelity highlighted only the element of promise. The physician professes his or her art (prudence) on behalf of someone (fidelity). The discussion barely adumbrated the moral advisories and the ritual/habitual routines that flow from this double profession and promise. But, in closing, I need to make explicit the element of gift that precedes and generates the power of this double promise to something on behalf of someone. Otherwise, we interpret the professional either contractually as the seller of services or contractually/philanthropically as a combined seller/giver of services but not covenantally as a person whose actions rest upon a comprehensive receiving. We interpret the physician as benefactor but overlook his or her deepest identity as a beneficiary. Thereby we fail to trace the three covenantal virtues to their original source and their final resource to which the virtues of gratitude and hope testify.

Before specifying the antecedent gifts implicit in the foregoing account of the three professional virtues, I must concede that one cannot fully appreciate the indebtedness of a human being by toting up the varying sacrifices and investments made by others to his or her benefit. The sense that one inexhaustibly receives presupposes a more transcendent source of donative activity than the sum of gifts received from others. The formula borrowed from Josef Pieper – "Being precedes Truth and . . . Truth precedes Goodness" [6] – neatly captures the element of transcendent gift upon which the virtues depend. Knowing depends upon the taking in of Being; and goodness (or doing aright) depends upon that receptive knowing. In the biblical tradition, this transcendent source is a Being which is a Being-true, which, in covenant fidelity, secretly gives root to every gift between human beings. The secondary gifts in the human order of giving and receiving can only (and imperfectly) signify this primary gift.

But giving at the human level also contributes to the physician's knowing and doing. The discussion of public-spiritedness has already made that clear. No one can graduate from a modern university and professional school and think of himself or herself as a self-made man or woman. Practicing physicians cannot survive a single day without drawing upon the massive research traditions of their profession that support whatever prowess is theirs in diagnosis, prognosis, and therapy. Not only past generations of researchers, but countless patients, many of them poor, have submitted to the experiments that now make the physician seem so smart in the office, the clinic, or the hospital. Further, that research tradition does not effectively connect with the individual patient unless the patient himself, presents, complains, lays bare his vulnerability and distress. The successful interview with the patient requires addressing, but also being addressed, giving, but also taking in, the tongue, but also the ear.

Thus, professionals who, in fidelity, benefit their patients do not do so as pure benefactors. Idealistic members of the helping professions like to define themselves by their giving or serving alone – with others indebted to them. A reciprocity, however, of giving and receiving is at work in the professional relationship that needs to be acknowledged. The physician has received abundantly from her patients, not only in all the aforementioned ways but also by their petitions for help which confer upon the physician her identity. Patients help give the professional her calling. In answering the call, the professional gets back nothing less than what she is – a remarkable gift indeed.

However, in trying to make good on their calling, professionals discover that they fall short. They promise a deep-going, core identity with their patients, but discover that drawing close to misery and deprivation threatens to suck them into a whirlpool. They risk getting mired down in the mess and confusion of their patients' lives. Some physicians respond to human affliction somewhat heroically and grimly, playing the role of savior, the anointed defender and guardian of the woe beset. Others, discouraged, recognize the very limited, partial, and temporary nature of the relief which they can offer; they burn out. Still others respond self-protectively, finding a way to shield themselves defensively from the stricken, radio-active patient. While differing, these varied responses share in common a metaphysical gloom; they raise the question of the place of the virtue of hope in the helping professions.

The virtue of hope does not enter into the picture if a covenantal ethic generates a moral principle of unconditional fidelity, but nothing more. Indeed, as a moral principle alone, it only intensifies professional burdens and increases the possibility of moral exhaustion and burnout. A covenantal ethic, however, depends upon the conviction that one's own promise-keeping rests, not simply on a moral principle of keeping one's promises, but upon an all-surrounding promissory event. God has made and will keep his promises to humankind.

This promise defines men and women not only retrospectively in gratitude but prospectively in hope. It comforts them in the midst of their professional excesses, shortfall, and burnout with the knowledge that God will not abandon his creatures, either those who give or those who receive care. God's promise gives to human promise-keeping some buoyancy. To the degree that professionals persist in "being-true" to their promises, even with some small measure of resiliency, they offer their troubled patients a tiny emblem of that ultimate hope.

Southern Methodist University
Dallas, Texas
U.S.A.

NOTE

[1] This essay draws on the three sources in my writings cited in the Bibliography, but it orders the thought in a way I have not heretofore proposed.

BIBLIOGRAPHY

1. Cassell, E.: 1979, *The Healer's Art*, Penguin Books.
2. Edelstein, L.: 1967, *Ancient Medicine, Selected Papers of Ludwig Edelstein*, Owsei Temkin, and C. Lilian Temkin (eds.), Johns Hopkins Press, Baltimore, MD.
3. May, W. F., 1983, *The Physician's Covenant*, Westminster Press, Philadelphia, PA, now in Louisville, KY
4. May, W. F.: 1984, 'The Virtues in a Professional Setting', in *Soundings*, vol. LXVII, No. 3 (Fall), 245–266.
5. May, W. F.: 1991, 'The Beleaguered Rulers: the Public Obligation of the Professional', the *Kennedy Institute of Ethics Journal*, vol. 2, No. 1, 25–41.
6. Pieper, J.: 1954, *The Four Cardinal Virtues*, University of Notre Dame Press, Notre Dame, Indiana.
7. Wolin, S.: 19??, *Politics and Vision*, Little, Brown and Company, Inc. Boston.

ED DUBOSE

TRUST IN THE CLINICAL ENCOUNTER: IMPLICATIONS FOR A COVENANT MODEL

1. INTRODUCTION

It is axiomatic that trust is essential to social relationships, certainly to the clinical encounter that lies at the heart of medical care. Trust appears most precarious, and most necessary, at those times when our vulnerability is the greatest. In health care, the primary relationship of the caregiver and the one seeking care acts to shield us from a stark confrontation with our finitude, from the feelings of helplessness and dislocation that occur when illness casts us out of our everyday life and deposits us in a different place ([18], p. 31).[1] The nature of illness and health care makes trust a basic ingredient in the clinical encounter between patients and clinicians, but changes in the cultural and social framework within which medical relationships exist, and of which they are a part, are affecting the way trust shapes the nature of that encounter.[2]

Because being ill creates dislocation and dependency, one is forced to trust the knowledge and skills of physicians and their desire to support rather than exploit his or her vulnerability. There are three elements supporting the medical profession's claim to the public's trust. Physicians, in essence, can be trusted with their knowledge and social dominance because:
1. they possess the formal knowledge and skill;
2. they act in our best interests; and
3. their commitment to one and two is inherent in the nature of their profession.

However, as Veatch argues, there are good reasons why professionals ought not to be able to know what their clients' best interests are. In addition, it is conceptually impossible, according to contemporary philosophy of science, to present "value-free facts" to the client. Finally, we cannot take for granted the traditional notion that professionals can be trusted to act on a univocal set of virtues inherent in the profession or that there is a single, definitive conceptualization of how the profession ought to be practiced ([49], pp. 160–161). If these traditional fiduciary claims associated with the medical profession are called into question, how meaningful in the clinical encounter is talk of trust when one can no longer assume that physicians naturally deserve it?

Consideration of this question is necessary before any attempt to revisit the theological covenant model for clinical ethics. In the last twenty-five years, we have moved from a positivist, rational, and problem solving focus in medicine to a sense that things are out of sorts. Medicine, like other American professions, has suffered a stunning loss of public trust since the 1970s. With

45

G.P. McKenny and J.R. Sande (eds.), Theological Analyses of the Clinical Encounter, 45–66.
© 1994 Kluwer Academic Publishers. Printed in the Netherlands.

relative suddenness medicine has become a focus for debate and ambiguity, if not skepticism and hostility about physicians' economic motives and dedication to their patients' interests ([5], p. 207). The time seems ripe for a review of the covenant model for new understandings of the clinical encounter. If the basic ingredient in clinical relationships can no longer be taken for granted, then the language of covenant as a basis for relationship is called into question, along with the secular language of contract. However, in the absence of trust, people seek a more direct control over their relationships, and at least on an immediate emotional level, contract language offers some measure of control, some way to limit the ambiguity of social interactions.

It is not coincidental that the sense of uneasiness in social relations corresponds with an increased attention to postmodern issues.[3] In the postmodern environment, I would argue, there is a pressure for control of the medical profession because of the deconstructionist critique of metaphysics, of the concepts of causality, identity, and truth. Where once one could assume that "the covenantal image of professional responsibility" represented trust, predictability, and accountability, in the postmodern clinical relationship emphasis has shifted to the image of contract, in an effort to ensure these characteristics ([7], p. 104). As a result, trust becomes a bargaining chip in relationships that develop to protect the parties, one from the other.

Therefore, the aim of this paper is to examine why attention to an erosion of trust must be included in any effort to revitalize a covenant model for the clinical encounter. First, it is important to recognize that the issue is not simply to overcome the problems caused by a secularization of medicine by restoring a covenantal focus in health care. Although a backlash has begun against deconstructionist thought, its attempt to expose the illusion of metaphysics (that behind, beneath, or within the play of appearances there is a truth which can be discovered, uncovered, or disclosed) serves to undercut the pretensions of secularization, as well as theology ([50], p. 114). If there are no foundational truths to serve as the basis of our relations with one another, we would seem to be left merely with relationships of mutual self-defense and self-interest. Second, looking more closely at the physician-patient relationship, I stress the important role trust plays in structuring fiduciary relationships. In this analysis, there is no decentering of the physician, but a decentering of the profession's fiduciary commitment to patients' best interests. Third, if the profession's traditional fiduciary claim is abridged, distrust becomes the functional equivalent of trust in clinical relationships. Finally, the piece concludes with a brief review of the implications of distrust in ascribing a covenantal basis to the clinical relationship.

2. THE PROBLEM OF TRUST IN A POSTMODERN ENVIRONMENT

The way in which trust operates or functions to create, sustain, and modify social relationships is a complex phenomenon that is difficult to define and distinguish. For the purposes of this discussion, trust is confidence in or reliance upon someone or something without investigation or evidence ([9], p. 180). Therefore, trust is always accompanied by an element of risk. In spite of the risk, or more accurately, because of it, the person or thing in which trust is reposed becomes a locus of belief, expectation, and hope.

Trust also represents a medium of social exchange which serves to regulate our relations with each other [33]. By reducing the complexity of social situations, especially as this complexity results from other people's freedom, trust shapes our expectations of others and of the situations in which we find ourselves. Ordinarily we assume that people will act consistently with the image or personality that they have made socially visible. However, as Barber argues, trust is not a function of individual personality variables nor of abstract moral argument, but a phenomenon of social and cultural variables ([1], p. 5). We may trust someone's knowledge or character, but our expectations of that person also are shaped by social institutions and cultural values. Therefore, broad social changes affect the meaning and interpretation of trust, with a corresponding effect on interpersonal relationships.[4]

Coupled with general concerns about rising health care costs, physicians now often are perceived as functioning "with the business ethic rather than the professional ethic" ([29], p. 2879). Also, rather than engendering trust, the success of technological medicine in the last generation often raises people's anxieties about their ability to make choices for themselves. Increased medical reporting raises public expectations of success in medical matters, yet it also exposes the dilemmas caused by medical progress. Our expectations of medicine are accompanied by a sense of the gap between professional and public knowledge, exacerbating anxieties about our vulnerability in clinical encounters. A demand for autonomy in decision-making contributes to an uneasiness with professional privilege and power. Public concern with controlling medical expenditures and professional power is redrawing the de facto secular "contract" between the medical profession and society, subjecting medical care to the discipline of politics or markets, or reorganizing its basic institutional structure ([45], p. 380).

The Secularization of Medicine

McCormick believes that these developments reflect the secularization of medicine, in which the profession becomes divorced from a moral tradition that formerly grounded the values that distinguished medicine as a human service. As a result, medicine is preoccupied with factors (insurance premiums, business atmosphere, accountability, competition, etc.) that are peripheral to and distract from care. Medical knowledge and skill are construed as commodities belonging to the physician, which can be dispensed on his or her own terms in the marketplace. The commodities model underscores a technocratic medical care that too easily focuses on the administration of things – with people among the chief things to be administered ([32], p. 1133). It emphasizes procedure and is guided by principles that give a privileged place to efficiency, problem solving, and beneficence. In bioethical reflection, the counterpart to this secularized model of medicine

holds that moral conduct consists of duties and rights determined by these rules. This secularized model, emphasizing social utility, is appropriate to govern interactions of strangers and of bureaucratic relationships within institutions and organizations ([43], p. 2).

As a result of these developments, there is an increasing uneasiness (a sense of dis-ease) within the clinical relationship which paradoxically demands the highest degrees of trust between the parties (a relationship . . .).

Too often conceptualizations of the clinical relationship presuppose a view of the person as independent and solitary, related to others by necessity. In this perception the individual becomes the only meaningful level of moral analysis. McCormick claims that it is now all but universally admitted in Western circles that individual decision-making regarding medical treatment is a necessary part of individual dignity. However, what is not so well recognized is that this heavy emphasis on the individual produces an "absolutization of autonomy" and an intolerance of dependence on others. Rejection of our own dependence, McCormick argues, means ultimately rejection of our interdependence ([32], p. 1132).

This rejection leads to a disregard for the complex social nature of trust in interpersonal relations. Trust is important only if our social and communal existence is taken seriously. Individualism treats social ethics as an afterthought because social arrangements are an afterthought of convenience and self-interest. Under these conditions, self-interest drives the clinical encounter towards an increasingly formal basis, with an emphasis on regulation and negotiated services between autonomous, "wary" parties, "treating each other as profit opportunities" ([31], p. 37). It is no wonder that the contract language of the marketplace has become common in health care. Dissatisfaction with an emphasis on individualism and principles revives attention to the covenant model as a basis for the relationships within the clinical encounter.

Theological understandings of bioethics emphasized a sense of community and enunciated an ethic of giving and receiving, of caring, of helping, and a medical moral responsibility premised on covenantal, interpersonal relations between physicians and patients, and between the medical profession and society ([43], p. 2). The covenant model for clinical relations is grounded in trust as an indispensable basis for relations between the parties. Ramsey, for example, wrote of reciprocal trust between doctors and patients grounded in a cooperative agreement of covenant-fidelity, as that which allows the healing relationship to occur [37]. According to him, the physician's authority is generated by the patient's free choice and by the mutual trust and faithfulness of the relationship. However, a generation later, with the turn towards individualism and the absolutization of autonomy in health care, a relationship of mutual trust and faithfulness seems the exception, rather than the rule.

McCormick's argument is suggestive, and if the secularization of medicine was the only problem, appeals to a theological reworking of the covenant model might reverse the movement to individualism and the absolutization of autonomy that he laments. However, there is another dimension underlying the erosion of our interdependence and our trust in each other. The influence of French poststructuralist thought, the most familiar example of which is the philosophical deconstructionism of Jacques Derrida, more deeply challenges our sense of interdependence and mutual trust. To explain this challenge, it is necessary to discuss Derrida's deconstructionist view of language and what he means by logocentrism. There will be traces of his views throughout the subsequent discussion.

Deconstruction and the Myth of Presence
Derrida's philosophy is wonderfully complex, but a good place to start unpacking his ideas is with Ferdinand de Saussure's language theory of signs. All signs, Saussure claims, are composed of two different aspects, which he designated as the *signifier* and the *signified*. The sound image made by the word "physician," for example, is the signifier, and the concept or meaning behind the word "physician" is the signified. The structural relationship between the two constitutes a linguistic *sign*, the combinations of which make up language. Both signifiers and signifieds have identities only as functional parts of systems of signifiers and signifieds (see [39], pp. 23–29). The signifier "physician" is the signifier only because it stands in opposition to the signifiers "patient," "nurse," etc. The signified of "physician" is the concept it is only because of the concepts it is not; the meaning we associate with physician is not the meaning associated with nurse, with patient, etc. In this theory of the sign, there is a correspondence between a signifier and its signified.

In Derrida's view of language, the signifier does not reflect a signified, as a mirror reflects an image. Instead, he is highly critical of the unity of the stable sign: there is no fixed distinction between signifier and signified; there is no one-to-one correspondence between propositions and reality. Signifiers and signifieds are continually breaking apart and reattaching in new combinations, thus overthrowing Saussure's model of the sign, according to which the signifier and the signified relate as if two sides of the same sheet of paper. Suppose one wanted to know the meaning of a signifier and went to the dictionary; all the person will find will be yet more signifiers, whose meaning one would have to look up, and so on. The process is not only infinite but circular: signifiers keep transforming into signifieds, and vice versa, and one never arrives at a final signified which is not a signifier in itself ([40], p. 35).

In other words, Derrida argues that meaning is not immediately present in a sign. Since the meaning of a sign is a matter of what the sign is "not," this meaning is always in some sense absent from it, too. Therefore, meaning becomes scattered along the whole chain of signifiers; it is never fully present in any one sign alone, but is rather a kind of flickering of presence and absence together. For Derrida, the structure of the sign is determined by the *trace* (implying a track or footprint) of that other which is forever absent (see [12], pp. 70–71).

The implication of this is that no meaning can ever be finally achieved. Sign will always lead to sign, no one can make the "means" (the sign) and the "end" (meaning) become identical: no experiment will produce final, univocal results; no diagnosis will summarize some "real" physical condition. Moreover, since a sign appears in different contexts, it is never absolutely the same. Meaning will never stay quite the same from context to context; the signified will be altered by the various chains of signifiers in which it is entangled. Therefore, no patient will be the same in different encounters; no two cases of pneumonia can be the same. In Derrida's view, language is a very unstable affair. It is an illusion for me to believe that I can ever be fully present to you in what I say or write, because to use signs at all entails my meaning being always somehow dispersed, divided, and never quite at one with myself.

Derrida's method of deconstructing language has been generalized to a "deconstruction of presence," an overturning of the philosophical assumption of an immediately available area of certainty or meaning that lies behind our words, in which we know experience with an immediate, unmediated clarity. Words can never describe reality; our experience is created by the words we use and the way we use them. The challenge that this deconstruction of presence presents to the clinical relationship will be discussed in section five. There is one more important component of Derrida's perception that affects any consideration of a covenant model for that relationship.

Since the time when the term "metaphysics" was first applied to the text that contained Aristotle's investigation of the essential nature of Being qua being, the term has been used to refer to such an investigation ([39], p. 5). In Western metaphysics, the study of Being qua being (ontology) is tied to theology, the investigation of the highest being which is the necessary condition for the possibility and actuality of all other beings. As a result, in deconstructionist eyes Western philosophy and theology has become "logocentric," committed to a belief in some ultimate "word," presence, essence, truth, or reality which will act as the foundation of all our thought, language, and experience. According to Derrida, Western philosophy seeks the sign that will give meaning to all others – the "transcendental signifier" – and to the unquestionable meaning to which all our signs can be seen to point (the "transcendental signified") ([12], p. 49). In theology, "God" is that sign, and attempts to develop a covenantal basis for social relations grounded in this sign become, for deconstructionists, logocentric fictions characteristic of the modern post-Enlightenment age.

To put this another way, postmodern critics attempt to subvert what they find to be the tyranny of comprehensive worldviews or visions of the *whole*, in relationship to which *parts* are to derive their meaning. Such visions of the whole are characterized by Jean-Francois Lyotard as "metanarratives" peculiar to modernity and its distinct aspirations ([30], p. 84). The idea of health as a condition to which we have access with the right biomedical knowledge and understanding represents such a modern metanarrative. The concept of "health care," in relation to which all of the institutions, models, roles, and values associated with that sign take their meaning, reflects a complex of relationships intended to correspond to this modern "reality" of health. Within this metanarrative, relationships are hierarchically arranged, inherent *power-plays*, which exclude, distort, and suppress any element which "threatens" them ([50], p. 114). The physician-patient relationship is an example of such a power-play. Physicians' appeals to biomedical authority reflect a narrative of mastery over natural, biophysiological processes; because of their power-full knowledge, physicians are able to act as gatekeepers to health. The professionalization of their knowledge-based skill becomes a power-play that subordinates patients in clinical relationships.

In sum, this postmodern style of criticism, influenced by deconstructionist thought, attacks any thought-system that depends on a first principle or unimpeachable ground upon which a whole hierarchy of meanings may be constructed. The quest for the "thing in itself" reveals a misapprehension: no final meaning can be fixed or decided upon; thus, interpretation abounds. The postmodern style is tinged by an undercurrent of suspicion: claims to univocal meaning, final truth, or logos are continually undermined, and we are left with the finitude of perspectivism. Postmodernism, then, is a/theological [46].

The argument is not that a theological covenant model cannot be applied to the clinical encounter, but that a postmodern critique of "presence" shifts the way in which trust operates in the relationship between patient and physician; theologically-based covenant models must take this shift into account. In short, there is a serious challenge to trust in the clinical context: what Pellegrino calls "an ethics of distrust," i.e., the formal attack on ". . . the very concept and possibility of trust relationships with professionals" [34], p. 79). Such an ethic would place tighter restrictions on physicians or eliminate the need for trust entirely. In the postmodern age of suspicion, trust cannot be assumed or taken for granted in our relations with professionals. Although trust always has been fragile in social relations, people now seem more conscious of the risks of trusting others. As a result, both physicians and patients are seeking protection and control, to reduce or eliminate the disruption in their lives represented by their respective encounters with illness and with each other. The implications for the covenant model are serious, if it is to be re-vitalized in clinical care.

3. THE COVENANT RELATIONSHIP AND THE PROFESSION'S FIDUCIARY COMMITMENT

It is an inescapable feature of the medical situation that people's ill-health, misfortune, and disadvantage are sources of status and income for others who offer their services as physicians. Because the patient's vulnerability makes him or her susceptible to abuse by the physician's self-interest, the relationship carries moral connotations. Society allows physicians to develop their knowledge and skills, and grants the doctor autonomy and social status, in return for the assurance that physicians can be trusted to use their knowledge only for the patient's benefit. Therefore, medicine maintains a public declaration that any physician will use his or her knowledge and skill to serve the client's needs. This commitment helps shape public expectations of the profession. Physicians are assumed to be competent, trustworthy, and client-centered. As a result, we approach the physician with expectations that allow us to be vulnerable.

These expectations are taken for granted by both physicians and patients. They do not stem from any personal benevolence or private virtue on the part of the physician, but from the very nature of the relationship between physician and patient. The patient places himself or herself in the hands of the doctor not as a friend to whom he has come for advice, but as the embodiment and agent of a certain kind of formal knowledge and skill placed in the patient's service ([44], p. 28). The social exchange between the two is based on trust, which serves in turn as a means of social control in regulating the exercise of power and as a mechanism of integration by creating and sustaining social solidarity. One model of this social exchange is the covenant.

The Covenant Relationship and the Medical Profession
In referring to a certain kind of relationship, a "covenant" among persons is a binding, enduring relationship of mutual loyalty or devotion. Theologically, within the Jewish and Christian traditions such a relationship is grounded in the relationship with God, with its rich connotations of trust and obligation ([38], p. 7). In scriptural references to covenant, an emphasis is placed on God's steadfast commitment: Israel can trust in God not because of the people's righteousness but solely because of God's covenantal guarantee of faithfulness to them. The Bible also sets out God's trustworthiness as a model for humans to follow; the covenant with God is the source and pattern of human covenants. Because of God's covenant with them, people are obligated in covenant to love, aid, and comfort each other ([13], p. 1136). Therefore, covenants imply an affirmation of each member in the relationship, a focus on the relationship between the members rather than on the stipulated obligations, and an emphasis upon mutual relatedness and enduring responsibility, even when the members are unfaithful or when one or the other is vulnerable to the other in the relationship.

According to Dennis Campbell, this bonding, when rightly understood, had practical behavioral consequences. "To be a true child of Israel was to act in certain ways which were predictable. . . . As a result, the covenant implied trust, predictability, and accountability" ([7], p. 104). These virtues have become central to the medical profession's identity: because of their special knowledge and power, physicians are pledged to do no harm and to act in their patients' best interests. Because of this covenantal declaration, physicians are not only able to perform what they have pledged to do, but they can be counted on to want to do so ([38], p. 12). Therefore, as Ramsey claimed, what characterizes the covenant relationship in health care is the emphasis on fidelity and trust as the basic means of exchange between physicians and their patients. The moral worth of the relationship does not depend upon the outcome of the relationship, but upon the immediate interaction of the doctor and patient. Trust becomes based on mutually and freely agreed upon terms, establishing obligations and responsibilities between the two parties. This covenant model dovetails nicely with the fiduciary claim that physicians are bound to service. From a sociological point of view, such a claim has enabled physicians to dominate the clinical relationship, either individually or collectively.

Trust and Professional Dominance
The perception that physicians are bound to service has encouraged the rise of medicine to professional status in the public eye. The idea of the professional is a structuring concept with far-reaching potential for social relationships. According to Eliot Freidson's professional dominance theory, the idea of profession represents a special kind of occupation and an avowal or promise (see [19, 20, 21]). As he writes

Unlike most occupations, [a profession] is autonomous or self-directing. The occupation sustains this special status by its persuasive profession of the extraordinary trustworthiness of its members. The trustworthiness it professes naturally includes ethicality and also knowledgeable skill. In fact, the profession claims to be the most reliable authority on the nature of the reality it deals with [18].

Because of the medical profession's possession of expert knowledge and skill, and its avowed service orientation, physicians enjoy a great deal of autonomy and control over their work.

In a technological, knowledge-driven society, professionals exert unprecedented power. Therefore, it is important to recognize the way in which power and authority structure professional-client relationships. As long as the public is convinced that physicians possess special knowledge and skill and can be trusted to act in the public's best interest, society grants the profession license to exercise its knowledge and control its standards of practice. It also confers upon them social prestige and financial reward. Although people ordinarily are resentful of the dominant party in an asymmetrical relationship, in relations with physicians distrust is suspended or bracketed: physicians' claim to fiduciary status is assumed until there is sufficient reason to doubt it. In this way, a dedication to fiduciary responsibility serves as a form of social control by a profession seeking power and prestige in its work [3]. At the same time, fiduciary responsibility also serves as a control or restriction on professional power. The public declaration to act in a patient's interests and to do no harm limits the physician's freedom to act irresponsibly by holding him or her accountable to professional standards.

Still, because these standards largely are controlled by its members, people are vulnerable to the medical profession's power. The recent emphasis on patient autonomy reflects an effort to counteract that professional power and the paternalism it can breed by redrawing the balance of power in the clinical relationship. For example, picking up on the covenant-fidelity model, Veatch argued that the mutual trust of the marriage contract defines the content of the clinical relationship, allowing the client and physician to pursue jointly developed goals, ensuring that both of their interests are served (see [48]). To close the power gap between a privileged, powerful class of health professionals and their vulnerable patients, in his contract model patients are seen as autonomous entities, encouraged to make independent, self-defining, and determining judgments about their medical care. As a result, health care becomes an item of negotiation between two powerful and autonomous entities in the medical relationship. To foster a balance of power in the clinical relationship, a sharp accent has been given to the patient's individualism, self-determination, and autonomy in the clinical context.[5]

These developments in clinical relations reflect a broader shift in the social framework that sustains our experience. Within the postmodern perspective, a profession's claims to autonomy in its work mask the self-interest of its members. In a postmodern critique of the medical profession, for example, rather than being actively involved and choosing agents, patients are acted on by the "power structures" that comprise health care's hierarchical social systems, centered around the physician ([17], p. 210). As patients encounter more and more technologically dominated, knowledge-driven systems of health care, a correlative desire for control over their experience strengthens the ideology of individualism *and* the use of rules, procedures, and contracts to regulate physicians' behavior. An emphasis on physician accountability, the turn to courts to resolve disputes, and the development of bioethics all suggest a social desire for restructuring the balance of power in clinical relations. In the midst of all this maneuvering, however, trust as positive reliance on another without investigation or evidence fades in the clinical relationship, to be replaced by distrust as a means of social control.

4. THE PROBLEMATIC NATURE OF DISTRUST AS SOCIAL CONTROL

Because of their professed commitment to patients' best interests, physicians are the locus of patients' expectations of assistance, concern, and confidence. As long as these expectations are met, trust is granted, the risks of the clinical relationship are tolerated, and the relationship is sustained. When these fiduciary expectations are violated, however, the imbalance of power between parties becomes clear and disturbing, and trust is abridged.

Barber argues that the logic and limits of trust suggest that when the balance of power between parties is disrupted, distrust becomes the functional alternative and complement to trust as a form of social control over the dominant party ([1], p. 9, see also [28]). When people are distrustful, they require more substantial constraints in their relations with others; they are more careful and deliberate in achieving some equation of power. As such, then, distrust is not necessarily the result of nor the cause of a breakdown in the social system. Both trust and distrust serve to reduce social complexity and anxiety; both serve to enable people to function in a precarious situation. Distrust, however, as a medium of social exchange, upsets the fiduciary expectations of the clinical relationship by exposing its asymmetrical nature and restructuring the basis on which the interaction occurs and the roles the parties play. If distrust is becoming the medium of social exchange between physicians and patients, how deep is the problem?

Evidence indicates that while physicians still rank highest among professions, there has been a decline in the public's trust in the medical profession ([29], p. 2879). One possible explanation for such a change is that public trust in all professions is declining; trust in the medical profession remains high only in relation to trust in other professions. A second possibility is that the social matrix within which trust arises is undergoing revision, and our expectations of physicians are changing. Given their need for help, people are forced to trust physicians, yet are uneasy doing so. As a result, they seek ways to augment their own control in the clinical relationship by stressing autonomy and self-determination.

In the first case, it is often meant either that self interest has so captured professionals' agenda that their clients can no longer approach them with the traditional moral expectation that client interests will be paramount, or that societal interests in cost containment, the advancement of research, or other agendas force the professional to sacrifice the client for these other goods. If this is the trouble with the fiduciary relationship, it is potentially fixable. We would need to make physicians more committed to their patients, educate them in the virtues and character consistent with medicine, or make clear morally those cases where conflicting interests are legitimate. All of these "first-order criticisms" of the fiduciary relation assume that it makes sense to believe that if physicians are of good character and are committed to patients, then trust can be restored to the relationship ([49], p. 159).

However, in the second instance, there is a more fundamental problem for fiduciary relationships. We have been arguing that the medical profession's declaration of fiduciary commitment – the promise of its members to be client-centered, not self-interested, moral agents committed to serving the best interests of their patients – enables physicians to develop their knowledge and apply their expertise in practice. In the postmodern milieu, there is a growing awareness that knowledge and fiduciary claims are historically and culturally situated, that conceptually it is impossible to separate facts from values, and that physicians' self-interest is hidden behind the myth that they always and only seek their patient's best interests ([2], p. 6). If physicians cannot achieve value neutrality in their presentation of "the facts," and if they cannot elicit trust in their interaction with patients by appeals to some general set of virtues inherent in the profession, it may be that the very concept of trusting a professional is no longer tenable. Under these conditions, any fidelity model of social relations will be radically affected.

Trusting Physicians with the Facts

Because there is a clear disparity in knowledge between physician and patient, those who emphasize patient autonomy and self-determination encourage physicians to present their patients with the information necessary for informed decisions regarding the person's treatment. Since patients may have difficulty understanding medical information, the physician must be trusted to make a good faith effort to explain and inform. However, can physicians be trusted to give a patient the objective, unbiased facts and options necessary for such decisions? Can a doctor have access to such facts, at all?

Inferring Patients' Interests from the Clinical Perspective

It is by now well known that we do not come to situations, issues, and actions *de novo*. Rather, we come with a distinctive way of seeing and describing our lives and the world, a way learned in particular communities as part of an ongoing tradition. Physicians participate in a profession, with its own historical tradition, its own specific cognitive style, and its own particular way of interpreting its experience. As members of this profession, they inculcate a particular mind-set or medical vision defined by a system of relevances that establish interests and priorities. As a result, physicians may find patients' symptoms significant only in the terms of explanatory empirical science. Given such a clinical perspective, which incorporates theory choice, selection of which observations are valuable, and choice of levels of significance in drawing conclusions, can a physician objectively reflect facts to someone who does not share the same frame of reference ([47], p. 223)? If people understand their experience from within different systems of meaning, there are no value-free facts, easily transferable, equally clear and univocal in any context. Each physician operates out of a professional frame of reference, usually accented by a particular specialty, which shapes the way in which the "facts" are constructed and presented to patients.

Physicians as Readers of Signs

The problem with knowing and communicating the facts runs deeper. Due to their training, physicians are less observers of discrete disease entities than intense readers of signs. The details that are noted by physicians are not attempts to record the world as it presents itself to their eyes but compilations of significant markers within their frame of reference (see [23], p. 86). Physicians are trained to see illness essentially as a collection of physical signs or symptoms marking particular diseases. Diagnosis in the medical encounter represents a "telling" of what these signs reveal. The doctor spends years collecting and studying signs, and this activity basically sets the pattern of his or her observations thereafter. Physicians see and interpret what they are trained to see ([10], p. 205); they look for the familiar.

However, in the deconstructionist account, meaning is not immediately present in a sign. For example, medical tests are designed for one or another purpose, and the results tell one thing but not another. Therefore, a physician's "telling" takes its value not by representing what happened; rather the telling of what happens in the clinical encounter also depends upon what did not happen, what is not told of what happened, and what cannot be told at all ([16], p. 197). A diagnosis depends not only on what the signs suggest, but also on what does not show up in the particular tests. David Hilfiker writes of the difficulty of convincing a patient that her headaches were the result of her lifestyle, not a brain tumor. Although there was every sign that the patient suffered from stress (the woman was severely overweight, smoked heavily, had high blood pressure, and a bad marriage), there is always a chance that a CAT scan would find a tumor ([25], pp. 50–51). Second, representing the facts to a patient or some other person involved in the clinical situation requires discrimination: what does the person want to hear, not want to hear, or need to hear. What is left out of the telling has an importance. Finally, what cannot be told at all may be the most important part of the telling. After all, medicine is a highly uncertain science; uncertainty is anxiety-producing for all concerned. Therefore, patient and physician long for what they both lack: the "presence" of certainty – a final, absolute understanding of the situation and a sure knowledge that doctors know what they are doing. As a result of this longing, medical tests and the physician's knowledge and "reading of the signs" are invested with an authority and power beyond their means. While we want to trust that knowledge, such trust cannot be grounded in the facts "presented" to us. Certain knowledge is always absent.

To the charge that professionals are unable to present the facts objectively, not because of their shortcomings as professionals but because of the inherent limits in the process of reporting any kind of knowledge, the liberal response has been that physicians should share their "best guesses," uncertainties, and possibilities with their patients, out of respect and with a commitment to patient autonomy ([49], p. 160, p. 164f.) If the physician is committed to the patient's best interests, it is argued, such respect for the person's autonomy and self-determination allows for a shared, mutual definition of these best interests. However, knowing a patient's "best interests" is impossible.

Trusting Physicians with Patients' Best Interests

According to the profession's fiduciary claims, because illness diminishes a patient's autonomy the physician has a special obligation to protect that person's best interests. With the "absolutization of autonomy," it is now well established in bioethics that it may be morally necessary at times to sacrifice the patient's best interests in order to respect his or her autonomy ([49], p. 161). Equally, there are times when serving the patient's best interests will conflict with that autonomy (e.g., when one is incompetent). This is not meant to imply that a "priestly" medical paternalism is justified or that physicians are "engineers" who follow the autonomous patient's orders ([48], pp. 5–6). Rather it indicates that communication between patient and physician is important in order to know where the patient places health within his or her hierarchy of goods. Pellegrino and Thomasma, for instance, argue that the physician works together with the patient "in trust" to develop a course of treatment most satisfactory to each party; i.e., a negotiated course of action which presumably would apply if the person becomes incompetent. Since health is a particular experience of a unique individual, a patient's best interests must be calculated with that person's help ([35], pp. 33–36). However, it is not clear that patients' are able to give clear accounts of their own interests ([42], p. 62). This problem is not only because patients change their calculation of best interest over time, but also because physicians and patients have differing interests that make it impossible to come to a clear and final agreement.

Physicians and Patients Have Different Interests
The world of everyday experience is made up of multiple concerns (physical, psychological, social, legal, economic, occupational, cultural, religious), all of which figure into a person's sense of his or her own well-being. Just as patients balance these concerns to calculate some sense of their own best interests, the physician also has a sense of his or her own best interests (which can include attention to accepted professional standards and values). Increasingly, physicians and patients are strangers, forced together by accident or by referral based on something far short of good knowledge of the patient's interests. Both parties may have a shared expectation of reaching a common agreement on clinical decisions, but their self-interests in the situation are different. Working from a biomedical view of reality, for example, physicians are interested in successfully recognizing and treating a disease (see [8], p. 75). Along with relief from pain and suffering, a patient is more concerned how one's condition affects his or her total well-being, composed of a complex mixture of physical, psychological, social, legal, economic, occupational, religious interests, among others. At most, the physician can know something of the patient's idea of well-being in the particular sphere of medical well-being, about which the physician is an expert ([49], p. 162).

However, even this claim that physicians are able only to know the patient's interest in the physician's sphere of expertise is problematical. The medical profession has many goals, objectives, or ends that have developed over time and serve to shape the concerns to which its members pay attention. As Veatch argues, in medicine the goals are variously preservation of life, cure of disease, relief of suffering, and promotion of health ([49], p. 162). These goals plausibly can conflict; what is more, different physicians absorb and reflect these goals or objectives differently. For example, according to professional standards, a doctor is not obligated to provide active euthanasia. However, one doctor may act to relieve a patient of suffering under certain circumstances, while another argues that assisted suicide devalues the profession's standards (see [36, 27]). In order to know what is in the patient's best interests, physicians need to know not only what various medical goals mean in the patient's sense of total well-being, but also the correct mixing and balancing of these conflicting goals within the medical sphere ([49], pp. 162–163). There is no unanimity within medicine about these goals.

There is one more argument on which the profession's trustworthiness is said to rest: even if physicians cannot present value-free facts, and if a fully common understanding of a patient's best interests is elusive, they can be trusted, ". . . if they are good practitioners, to manifest the virtues inherent in this profession." In this view, trusting physicians to be virtuous rests on the presumption that there is such a thing as a virtue inherent in medicine that physicians can be expected to know and exhibit to the general public ([49], p. 166).

Appeals to Physicians' Virtue

Because of their competence and fiduciary commitment, it is usually assumed that a physician can be trusted to serve the interests of patients, whether or not he or she knows the particular patient's interests. Rather than ask whether physicians are able to know their patients' interests at all, cannot physicians be trusted to manifest the virtues inherent in the profession?

Few people agree on the characteristics which determine professional status. Historically, a profession is an occupation for which one is paid for specialized skill and expertise. In the last century plumbers, dry-cleaners, and baseball players are among the many occupations which describe themselves as professionals, yet few would grant them the same respect as physicians. If people regard a particular occupation as a "status" profession, it is often assumed that there must be a fundamental quality of that group which sets it apart from occupational professions.[6] Thus, as status professionals physicians would manifest the particular virtues inherent in the medical profession. However, virtues are historical conceptualizations, growing out of underlying belief and value systems integrally related to a particular time and place. These conceptualizations, these systems, change over time. Can we determine, once and for all, what characterizes the "good" physician?

Appeals to Profession-specific Virtues
Some people claim that virtues are profession-specific. This view assumes that there is such a thing as a virtue inherent in the profession that the members can be expected to know and publicly express ([21], pp. 194–200). This idea assumes a stable institution, an organized community of shared identity in which acts and relations retain their identity in a variety of situations, over

time. However, can the medical profession determine the virtues that are appropriate to an ideal physician? The meaning of "the medical profession" and the dimensions of its work result from complex negotiations within a larger social matrix and vary over time [21]. The claim that we can deduce a list of essential virtues from an idealized medical profession ignores the historical particularity of the profession, the varied interests of its members, and its interaction with the larger society.

We might say that physicians are characterized by the function they serve in society, and that this function or role is consistent over time and in various situations. Particular virtues may vary over time, depending on how efficacious they are in supporting or structuring this function. For example, because of their specialized knowledge and skills, physicians act as gatekeepers to health care. Their commitment to knowledge and objectivity elicits the public's trust in the profession, enabling physicians to fulfill their function of maintaining social equilibrium by controlling access to and removal from the sick role [33]. The traits which support the function become associated with the profession and avowed by its members.

The gatekeeper role contains an implicit explanatory theory for the institutionalization of medicine's normative structure, with roots in industrial nations influenced by Anglo-American institutions. Similar to other social institutions, medicine develops, changes, and adapts historically in response to social, economic, and political changes. Accordingly, the medical profession is an occupation which has most effectively gained and works to maintain a monopoly in its market ([3], pp. 51–59). To gain control of its work, the medical profession developed an organized structure, an institutional ideology, limited access to its ranks, and claimed a "status" as a profession acting in its client's best interests. Professional virtues, then, are those traits which underscore the profession's power, reinforce its autonomy, and enable physicians to control their work in society. The historical nature of these virtues is not as important as the way in which they are used to realize professional power and authority.

Thus, it is a mistake to assume that there is a univocal concept of the medical profession. Rather, recognizing the historical nature of professional roles, and the individual nature of a physician's interpretations of his or her role, there are as many different conceptions characterizing the traits of the good physician as there are underlying systems of belief and value that can imagine professionals working in the medical sphere. "A Talmudic or feminist or Hippocratic or Christian conception of 'physicianing' may have so little to do with each other that they are essentially different professional roles" ([49], p. 167). Moreover, because there are many different conceptualizations of how a particular profession ought to be practiced, it also is a mistake to assume that a common function, with common traits, can be abstracted or distilled from that profession ([21], pp. 29–36). There can be no single set of virtues inherent in the medical profession.

5. THE CHALLENGE OF DISTRUST IN THE CLINICAL ENCOUNTER

The argument about the shifting nature of trust in the postmodern environment is not to say that there is no trust or trustworthiness between people,
nor that discussion of the covenant model for clinical relations is pointless.
However, if the "taken for granted" foundations for trust between physicians
and patients are called into question, the disparities in social power are
revealed, and relations become unsettled as the participants seek a new measure
of control. According to a postmodern critique, this quest for power is
triggered by the desire for self-possession and control, to gain the security
in the relationship that one feels is lacking. Therefore, the erosion of
trust moves people away from a sense of interdependence to a desire for
independence (see [11], p. 91), and distrust begins to operate within the
relationship to control the interactions between the parties. In this postmodern
environment, there are two points to consider if the covenant model is to restore
a positive dimension of trust in the clinical relationship:
1. there are no metanarratives to serve as models of presence. Therefore, an
    emphasis on concepts such as covenant or health reveals the human desire
    to posit a central presence in experience; and
2. the "signs" of the modern medical relationship only mask the self-interest
    of the parties.
   First, Jews and Christians are part of religious traditions shaped by the
idea of the divine covenant as a model for social relations. The covenant stories
of these traditions give meaning to people's experience: they reflect the effort
of two persons or groups of people to recognize each other's concerns and
arrive at a mutually acceptable solution to the issue, regardless of its terms and
consequences, that lives out and articulates the relation of God and humankind
([38], p. 9). In the postmodern period, however, deconstructionists argue that
this covenant grounded in the "presence" of God is no longer anchored in
the history and traditions of the societal community. There are no metanarratives in the postmodern age [30]; there is no original Sinai or Easter
story from which covenant relationships draw a univocal meaning applicable
to all relationships.
   In the deconstructionist account, the covenant story is based on an absence;
there is no truth "behind" or "within" a text. Therefore, the "I am" of Mt. Sinai
is absent; the tomb is empty. In a desire for metaphysical presence, we want
to inform the "I am" with a particularity which sets it apart as unified – a single
protagonist – as different from others, invested with absolute priority and
privilege (see [46], pp. 36–37). We then assume that the covenant somehow
makes the relationship between two such particulars a unique one, when what
we really desire is for the relationship, for our experience, to be predictable.
For example, we want to take it for granted that a physician and a patient
are concerned, dependable, and knowledgeable about the other's unique
condition, committed to the other's best interests – each in his or her own
way trustworthy. However, social relations are contaminated by the patterns

of the subject's desire for the structure of presence. More and more, we tend to enter the clinical relationship suspecting the worst by demanding a certainty that medicine cannot provide. As Hauerwas has argued, ". . . modern medicine must necessarily fail, because a success commensurate to our desire is impossible" ([24], p. 202).

Second, in the desire for health and wholeness (the desire to preserve the "I am"), the physician is invested with power, privilege, and authority to which the patient is subordinated. However, in a sense the physician is only a token of what the patient lacks (charts, white coat, technical armamenture, process and procedure – all indicate the absence of health, the lack of wholeness). For a patient the physician represents the "sign" of health, a link with one's desire for the remembered, original condition prior to the disruption caused by the patient's illness. However, as Eco argues, such "signs" (e.g., the Hippocratic oath, the CAT scan, or a white lab coat) deceive by claiming to correspond with the promise of health (see [14], pp. 3–11). If physicians cannot know the facts and cannot be trusted to know their patients' best interests, they should admit it. Of course, it then becomes harder to risk trusting them with one's well-being.

It is no wonder that medical care is now seen as a commodity, spoken of in economic terms, detached from a moral tradition of human service. Such care, with an emphasis on and commitment to the controls of cost-containment and utility, has become homogenized; such care can be delivered anywhere, and thus is nowhere. Physicians and patients are characterized as "health care providers and consumers," the impersonality of whom is reflected in terms that bear no traces of any particular person. As clinical care increasingly is regulated, the commodity of care comes in one-size that really fits no one.

As a result, to create some sense of secure relationship, a concern for (dis)trust creeps into the de facto contract between providers and the public. To hide from the failure to be fully, finally autonomous, physicians and patients make trust a commodity to be bartered in exchange for a sense of security. Contract discourse offers a measure of control because physicians and patients seek predictability in their relations with each other. For example, the fear of prolonged dying, of one's life being out of one's control, drives the increasing emphasis on advanced directives or durable powers of attorney in health care. Patients desire control over what happens to them (via control over others), even from their comatose condition. Knowing our own or another's best interests is problematical, so we seek to bind each other to our will. In this sense, the attitude of distrust is itself a functional equivalent of trust: it does create relationships of a sort.

6. CONCLUSION

If postmodernism can be described as a culture of ontological doubt, the key shift from modernism has been the replacement of the plurality of interpretation by a sense that any "reality" is as meaningful or meaningless as any other. Within bioethics, the modern secularized sensibility produces a polarization between individualism and autonomy, on the one hand, and communitarianism and beneficence, on the other. The indeterminate nature of interpretation for bioethics (that there is no way to establish precedence among moral viewpoints) has left us with a sense of moral drift in medical care and leads to a profusion of bioethical theories (e.g., casuistry, narrative, principles, hermeneutics, and phenomenology). The notion that the postmodern sensibility involves a shift of emphasis from epistemology to ontology, a *deprivileging* turn from knowledge to experience, from theory to practice, from mind to body ([4], p. 281) has serious implications for trust in the clinical encounter.

One cannot assume that physicians can be trusted to know their client's best interests, present facts and options to them objectively, or exhibit virtues inherent in their professional roles. Likewise, patients cannot be trusted to know their own best interests, present facts objectively, or reflect a particular set of virtues. This does not imply that physicians or patients are "untrustworthy" in the sense that they are lacking in integrity or good character. It does not even imply that they are inherently self-serving or unable to serve interests of others before their own. But as long as the clinical relationship is rooted in a longing for a conflict-free, unambiguous "partnership," it will include a dimension of distrust as control. Under these conditions, distrust remains the basis for clinical relations in which each party seeks to absolutize his or her autonomy and power in the relationship.

This is not to say that the clinical relationship always is fated to be a power struggle of "us" versus "them." A deconstructionist critique applied to the clinical relationship represents a challenge and an opportunity to the theological covenant model. It challenges theologians to think in a radical way about the ambiguous manner in which Being both presents and hides itself in language, and about the tendency in modern Western culture (religious and secular) to aim at control and domination of the world and other people. By challenging the metanarrative presumptions of modern secular reductionism and instrumental rationality ([50], p. 115), as well as the particular reductionism of theological metanarratives, such a critique also opens more positive possibilities for a renewal of covenant language in clinical matters.

In seeking these possibilities, we must shift the rules of the search. Rather than seeking a covenantal relationship with others that somehow mirrors relationship with the Being of all beings, we must seek relationship with something like the Being of that set of beings that *I* encounter as *I* experience the world. This does not refer to some sort of ideal "I," some transcendental signifier that corresponds to each and every person. Such an approach reflects the metaphysics of presence we have been criticizing; it reinforces the belief that only *some* roads lead to a covenantal relationship – that, for example, those individuals who use religious language have a privileged access to meaning. That approach will only reinforce the movement to individualism and the absolutization of autonomy; it will encourage the use of trust as control in relationships. It may be that all of us, each and every one, have access through our everyday experience to covenantal relations. We hurt, we laugh, we affirm life as worth living and orient our lives around a host of interests expressed in our relations with others. The language of physicians (and theologians) can look down on all that, and not join in it and become part of it ([26], p. 222). It may be that physicians (and theologians) must set aside to some degree their rational objectivity, work in the neighborhood clinic, and give up the expectation that life finally can be healed, as if it can be restored to some proper attunement with the reality it inhabits.

The Park Ridge Center for the Study of Health, Faith and Ethics
Chicago, Illinois
U.S.A.

NOTES

[1] The patient-physician relationship is only one of a set of morally significant relationships in the clinical context. In this essay, I am interested in the effect of unsettled trust specifically on the clinical relationship between patient and physician.
[2] Trust is not a property of individuals but is achieved interpersonally within a particular social and cultural milieu. The idea of trust adheres to an individual no more than the idea of sickness does; both are media of interaction. When we trust someone or when someone is sick, our interactions with that person take on a certain tone. It is in this sense that trust or its lack functions as a medium of social exchange. (See [17], pp. 205–206.)
[3] In this essay, "postmodern" refers to a period of time, the present and future, and to a "style" of criticism. With regard to medicine, the postmodern period is characterized by the increasing effectiveness of medical technology, but also its increasing cost, and in response to that cost, increased government and corporate interventions, and the growing public awareness of scarce resources for care. As a style, "[p]ostmodernism is defined in the continual deconstruction of its definitions" ([17], p. 215). Deconstructionism can be described as "a philosophy that represents postmodernism, a period of time that follows the atrophy of modernity's various projects in science, ethics, aesthetics, and theology, and declares the fatal collapse of the rationalistic aspirations of the Enlightenment" ([50], p. 115).
[4] For example, the broad increase in patient self-determination can cause the medical profession to revise its code of professional ethics and encourage individual physicians to "share" decisions with their patients.
[5] Since contracts are negotiated between equals or near equals, this model obscures the clinical realities of serious illness and disease, and their impact on persons. It is difficult to envision someone who is ill, worried, and anxious, as an equal in the face of the physician's knowledge and skill ([34], p. 76; [35], p. 103).
[6] A "status" profession is one which is given qualitatively higher social status because it is assumed that financial reward is a secondary concern to its members. "Occupational" professions are those groups whose members use their knowledge and skill primarily to earn a living ([15], pp. 14, 32).

BIBLIOGRAPHY

1. Barber, B.: 1983, *The Logic and Limits of Trust*, Rutgers University Press, Rutgers.
2. Bartholome, W. G.: 1992, 'A Revolution in Understanding: How Ethics Has Transformed Health Care Decision Making', *Quality Review Bulletin* 18 (January), 6–11.
3. Berlant, J. L.: 1975, *Profession and Monopoly*, University of California Press, Berkeley.
4. Boyne, R.: 1988, 'The Art of the Body in the Discourse of Postmodernity', in M. Featherstone, M. Hepworth, and B. S. Turner (eds.), 1991, *The Body: Social Process and Cultural Theory*, Sage Publications, London, pp. 281–296.
5. Brandt, A. M.: 1991, 'Emerging Themes in the History of Medicine', *The Milbank Quarterly* 69(2), 199–214.
6. Brock, D. W.: 1989, 'Facts and Values in the Physician-Patient Relationship', in E. D. Pellegrino, R. M. Veatch, and J. P. Langan (eds.), *Ethics, Trust, and the Professions: Philosophical and Cultural Aspects*, Georgetown University Press, Washington, D.C., pp. 113–138.
7. Campbell, D. M.: 1982, *Doctors, Lawyers, Ministers: Christian Ethics in Professional Practice*, Abingdon, Nashville.
8. Churchill, L. R., and Churchill, S. W.: 1982, 'Storytelling in Medical Arenas: The Art of Self-determination', *Literature and Medicine* 1, 73–79.
9. Churchill, L. R.: 1989, 'Trust, Autonomy, and Advance Directives', *Journal of Religion and Health* 28, 175–183.
10. Daniel, S. L.: 1986, 'The Patient as Text: A Model of Clinical Hermeneutics', *Theoretical Medicine* 7, 195–210.
11. Derrida, J.: 1978, *Writing and Difference*, A. Bass (trans.), University of Chicago Press, Chicago.
12. Derrida, J.: 1976, *Of Grammatology*, G. C. Spivak (trans.), Johns Hopkins University Press, Baltimore.
13. Duntley, M. A.: 1991, 'Covenantal Ethics and Care for the Dying', *The Christian Century* 108, 1135–1137.
14. Eco, U.: 1985, 'Strategies of Lying', in M. Blonsky (ed.), *On Signs*, Beacon Press, Boston, pp. 3–11.
15. Elliot, P.: 1972, *The Sociology of the Professions*, Macmillan, London.
16. Frank, A. W.: 1990, 'The Self at the Funeral: An Ethnography on the Limits of Postmodernism', *Studies in Symbolic Interaction* 11, 191–206.
17. Frank, A. W.: 1991a, 'From Sick Role to Health Role: Deconstructing Parsons', in R. Robertson, and B. S. Turner (eds.), *Talcott Parsons: Theorist of Modernity*, Sage, London, pp. 205–216.
18. Frank, A. W.: 1991b, *At the Will of the Body*, Houghton Mifflin, Boston.
19. Freidson, E.: 1970, *Profession of Medicine: A Study in the Sociology of Applied Knowledge*, Dodd, Mead, New York.
20. Freidson, E.: 1975, *Doctoring Together: A Study of Professional Social Control*, Elsevier, New York.
21. Freidson, E.: 1986, *Professional Powers: A Study of the Institutionalization of Formal Knowledge*, University of Chicago Press, Chicago.
22. Goode, W. J.: 1957, 'Community Within a Community', *American Sociological Review* 22 (April), 194–200.
23. Greenblatt, S.: 1991, *Marvelous Possessions: The Wonder of the New World*, University of Chicago Press, Chicago.
24. Hauerwas, S., Bondi, R., and Burrell, D. B.: 1977, 'Medicine as a Tragic Profession', in *Truthfulness and Tragedy*, University of Notre Dame Press, London, pp. 184–202.
25. Hilfiker, D.: 1985, *Healing the Wounds*, Pantheon, New York.
26. Hill, R. E.: 1987, 'The Future of Ontotheology', in H. Ruf (ed.), *Religion, Ontotheology, and Deconstruction*, Paragon House, New York, pp. 211–226.

27. Kass, L. R.: 1991, 'Why Doctors Must Not Kill', *Commonweal*, Special Supplement (August 9), pp. 462–476.
28. Luhman, N.: 1988, 'Familiarity, Confidence, and Trust: Problems and Alternatives', in D. Gambetta (ed.), *Trust: Making and Breaking Cooperative Relations*, Basil Blackwell, London, pp. 94–108.
29. Lundberg, G.: 1985, 'Medicine – A Profession in Trouble?', *The Journal of the American Medical Association* 253 (May 17), 2879–2880.
30. Lyotard, J.: 1984, *The Postmodern Condition: A Report on Knowledge*, University of Minneapolis Press, Minneapolis.
31. May, W. F.: 1992, 'The Beleaguered Rulers: The Public Obligation of the Professional', *Kennedy Institute of Ethics Journal* 2, 25–41.
32. McCormick, R. A.: 1991, 'Physician-assisted Suicide: Flight from Compassion', *The Christian Century* 108 (December) 8, 1132–1134.
33. Parsons, T.: 1969, *Politics and Social Structure*, Free Press, New York.
34. Pellegrino, E. D.: 1991, 'Trust and Distrust in Professional Ethics', in E. D. Pellegrino, R. M. Veatch, and J. Plangan (eds.) *Ethics, Trust, and the Professions: Philosophical and Cultural Aspects*, Georgetown University Press, Washington, D.C., pp. 69–89.
35. Pellegrino, E. D., and Thomasma, D. C.: 1988, *For the Patient's Good*, Oxford University Press, New York.
36. Quill, T. E.: 1991, 'Death and Dignity: A Case of Individualized Decision Making', *The New England Journal of Medicine* 324 (March 7), 691–694.
37. Ramsey, P.: 1970, *The Patient as Person*, Yale University Press, New Haven.
38. Roehrs, W. R.: 1988, 'Divine Covenants: Their Structure and Function', *Concordia Journal* (January), 7–27.
39. Ruf, H.: 1987, 'The Origin of the Debate over Ontotheology and Deconstruction in the Texts of Wittgenstein and Derrida', in H. Ruf (ed.), *Religion, Ontotheology, and Deconstruction*, Paragon House, New York, pp. 3–42.
40. Sarup, M.: 1989, *Post-structuralism and Postmodernism*, The University of Georgia Press, Athens.
41. Schneidau, H. N.: 1976, *Sacred Discontent: The Bible and Western Tradition*, University of California Press, Berkeley.
42. Sehgal, A., Galbraith, A., Chesney, M., et al.: 1992, 'How Strictly Do Dialysis Patients Want Their Advance Directives Followed?', *The Journal of the American Medical Association* 267 (January 1), 59–63.
43. Siegler, M.: 1991, 'The Secularization of Medical Ethics', *Update* 7 (June), 1–2, 6–8.
44. Sokolowski, R.: 1991, 'The Fiduciary Relationship and the Nature of Professions', in E. D. Pellegrino, R. M. Veatch, and J. Plangan (eds.), *Ethics, Trust, and the Professions: Philosophical and Cultural Aspects*, Georgetown University Press, Washington, D.C., pp. 23–39.
45. Starr, P.: 1982, *The Social Transformation of American Medicine*, Basic Books, New York.
46. Taylor, M. G.: 1984, *Erring: A Post Modern Atheology, Part One*, University of Chicago Press, Chicago.
47. Toombs, S. K.: 1987, 'The Meaning of Illness: A Phenomenological Approach to the Patient-Physician Relationship', *The Journal of Medicine and Philosophy* 12, 219–240.
48. Veatch, R. M.: 1972, 'Models for Ethical Medicine in a Revolutionary Age', *Hastings Center Report* 2 (June), 5–7.
49. Veatch, R. M.: 1991, 'Is Trust of Professionals a Coherent Concept?', in E. D. Pellegrino, R. M. Veatch, and J. Plangan (eds.), *Ethics, Trust, and the Professions: Philosophical and Cultural Aspects*, Georgetown University Press, Washington, D.C., pp. 159–169.
50. Walsh, T. G.: 1989, 'Deconstruction, Countersecularization, and Communicative Action: Prelude to Metaphysics', in H. Ruf (ed.), *Religion, Ontotheology, and Deconstruction*, Paragon House, New York, pp. 114–126.

SECTION II

PRINCIPLES IN REVISION

DAVID C. THOMASMA AND EDMUND D. PELLEGRINO

AUTONOMY AND TRUST IN THE CLINICAL ENCOUNTER: REFLECTIONS FROM A THEOLOGICAL PERSPECTIVE

INTRODUCTION

The most distinctive feature of medical ethics in the last two decades has been the shift in the moral center of the clinical encounter from the physician to the patient. Centuries old benign paternalism has been displaced by patient autonomy. On the whole, this has been salubrious and physician-patient relationships as a result are more open, adult, and morally defensible than ever before.

But, as so often happens when sharp conceptual shifts occur, one or the other of the extremes of an antinomy tends to be absolutized. Many physicians and some ethicists take this to be the case with the principle of autonomy. Beneficence has erroneously been identified with paternalism. Autonomy has become the trump card when prima facie principles are in conflict. This dominance of autonomy is, in our view, in need of tempering by a more modulated balancing with beneficence and trust.

The principle of autonomy and its current dominance in moral discourse are largely products of an intensive philosophical inquiry into medical ethics. This is the first era in which philosophers have taken explicit interest in the clinical encounter. Heretofore, medical morals consisted largely of moral assertions by physicians about what they conceived their duties to be. For the first time in its history, the ethics of clinical medicine is receiving the critical philosophical inquiry it deserves.

For many patients and physicians, however, philosophical inquiry fails to engage some of the most important features of the clinical encounter. Philosophy by its very nature cannot admit sources of moral insight provided by a faith commitment. For many people, religion is still the ultimate source of morality, affecting every aspect of daily life. There is need, therefore, of more extensive theological reflection on the day-by-day realities of the way physicians and patients relate to one another than has been the case.

In this essay, we shall offer some theological reflections on the central and now dominant principle of autonomy and its relationship to trust. We hope to provide some corrective to the absolutization of autonomy at least for those who share a Christian commitment. On our view, the restoration of trust as an element of the clinical encounter is essential if patient autonomy is not to be self-defeating.

We do not wish to depreciate the value of increased patient autonomy. It is essential in a democratic, pluralistic, and technological society. It is indispensable to protection of the human rights of the sick. It is a safeguard against

69

G.P. McKenny and J.R. Sande (eds.), Theological Analyses of the Clinical Encounter, 69–84.
© 1994 *Kluwer Academic Publishers. Printed in the Netherlands.*

the imposition on the sick person of the values of technology, the doctor, the hospital, or the community. Protection of the patient's autonomy is a buffer against the most egregious moral insults inherent in today's impersonal clinical environments.

But, its ethical necessity notwithstanding, over-emphasis on autonomy to the exclusion of other principles and moral values has its negative aspects. It can unjustly jeopardize the moral agency of physicians, encumber the resources of society, or lead to physicians' dissociation from their patients' decisions. The resulting moral atomization is damaging to any sense of community and exacerbates the fact that we are already increasingly strangers to each other, unaware of each other's values, life histories, goals, and aspirations. Much of the move in the direction of autonomy was set in motion by increasing distrust of the technical, social, economic, and professional power of physicians. As autonomy has grown, trust has been reduced increasingly to a phenomenological illusion or to a logically incoherent concept. Autonomy and trust appear to be in a morally inverse relationship.

Our purpose in this essay is to examine the limits of the inverse relationship, from the standpoint of our own theory of beneficence in trust as it is illuminated by a faith commitment. Our central question is this: Can a theological perspective help us put the principle of autonomy into its proper perspective within a trust relationship both for the patient and the physician – and, if so, how may it do so?

In our inquiry, we will take a Christian perspective, or, more specifically, a Roman Catholic viewpoint, but one consistent with other Christian faith affirmations and with the Jewish tradition as well.

RESPECT FOR PERSONS IN RELIGION AND MEDICINE

The principle of autonomy rests on the deeper principle of respect for persons. Respect for the unique dignity of the human person is a convergence point at which medicine and religion reinforce one another. Both start with an understanding of the fragility and finitude of human life. Both affirm that health and happiness are appropriate ends and aims for human beings, and that altruism is a necessary virtue. Both recognize that a violation of a patient's autonomy is a maleficent act that vitiates the healing purpose of the clinical encounter. It is perhaps for those reasons that the earliest priests were physicians, that medicine still retains something of the hieratic, and that medicine and religion begin with solicitude for the predicament of human illness.

In their beginnings, religion and medicine were virtually one. They were embodied in the person of the priest/medicine-man/shaman who acted as mediator between the sick person and the forces of nature or the spirits presumed to be the causes of illness. The Babylonians and the friends of Job, for example, held illness to be a direct consequence of sin. Healing consisted in confession of sin or infractions against the community mores, e.g., violations of taboos, disrespect for sacred places, and the like. The physician healed the body by healing the spiritual ill that caused the disease. He intervened with God, the spirit world, or the forces of nature to remove the cause of illness and restore the sick person to the community. This primary mediator function was retained by the priests, more than physicians, until the rise of science in the 17th century.

Around 2,500 years ago, medicine began to separate itself from religion (*On the Art* in *Hippocrates II*, Loeb Library, Trans. W.H.S. Jones, pp. 185–218). Gradually, the physician became an independent healer. From the beginning, some physicians appreciated the special moral nature of the healing enterprise. They imposed upon themselves a set of obligations that had as a central theme the promotion of the welfare of the sick person. Beneficence and respect for the sick are at the heart of the ethical codes of the Indian, Chinese, and Greek physician. The Hippocratic Oath clearly acknowledged these obligations in its solemn promise to respect the patient's confidentiality and his or her physical and sexual vulnerability.

The genesis of respect for persons in religion is somewhat different. In its beginnings, religion emphasized respect for the members of one's own tribe, but not for those outside the tribe or nation, i.e., it focussed on kinship piety, and group solidarity. This more narrow awareness began to change around 500 B.C. with the prophetic movement in Judaism and Zoroastrianism and the development of Confucian philosophy in China and Buddhism in India. Some notion of the common bonds among humans and the universality of the experience of illness began to emerge.[1] No doubt this tendency to universalism occurred parallel to political events, organization, and greater international communication. Medicine, itself, responded to this common vision at about the same time.

At the heart of the prophetic movement of Judaism was the belief that a transcendent power made all human beings and called all of them to obey Him and be redeemed. A vision of God as a god beyond the boundaries of any one nation meant He could be worshipped wherever one was. This worship was to take the form of commitments to other human beings. Exclusivism and tribal loyalty are not compatible with this conviction. They are replaced with a belief in the inherent value of all human beings, a belief that demands respect for all persons. This view was central to the Christian worldview as well. It also corresponds with the universalism of Roman Stoicism and the Roman view of natural rights.

These two streams – respect for the patient in professional ethics and prophetic universalism – came together in the early Christian era. Pedro Lain-Engralgo [14] describes this relationship throughout the history of Western medicine. To be sure, the philosophical basis for respect for persons differs in Greek, Roman, Christian, and Enlightenment worldviews.[2] But, in each worldview, there is an underlying conviction of the worth and dignity of every person that coincides well with the Judeao-Christian religious view of all humans as children of the same God, linked as brother and sister and obliged to care for one another.

The antagonism some ethicists see between autonomy and beneficence can be mitigated by a theological medical ethic.[3] This is not to justify medical paternalism, which is too often confused with beneficence, but to assert that respect for persons is, in itself, a requirement of beneficence. For the Christian, this beneficence is necessary to the virtue of charity, i.e., the unselfish love the Christian gospels tell us we must show to our fellow human beings. Each human being must be free because each has worth, each is accountable to God. Each must, therefore, be free to follow his or her conscience in moral choices – medical or otherwise [1].

Viewed from a Christian perspective, however, autonomy is not absolute. The Christian is obliged to use his or her God-given freedom wisely and well. Autonomy is a necessary means to doing the right and the good, to fulfilling the stewardship of our own health. This means refraining from self-destruction by suicide or deleterious life-styles, or neglecting needed and appropriate medical care.

Here it is instructive to note the difference between a pure autonomy based argument for refusal of therapy and one that is based on a religious sense of obligation to God to be a steward of one's own life. In the former analysis, individuals are complete masters of their own bodies. Due to the technological interventions of modern medicine, this mastery is often placed at risk. One has a counterbalancing freedom to refuse all therapy in an effort to control the power of modern medicine over one's own self-determined actions.

By contrast, the obligations towards one's own freedom and one's body from a religious perspective are different. One must care for the body and the self, since they both are creations and come from a Higher Power to which they owe their existence. Consequently, one is not free to refuse all therapy if it could reasonably be used to saved one's life. If the treatment were not considered physically or economically burdensome, and it had a reasonable chance of offering some benefit, a Christian would be obliged to accept the treatment or be resigned to its use. There are important caveats to both the autonomy-based view of the body and the religious view. Refusal of care is one important place where a religious perspective on the value of life and its origins makes a difference in the way one reasons about medical ethics [16].

But, if a patient refuses to acknowledge these duties, the physician cannot impose them upon him or her. Strong paternalism is uncharitable because freedom to choose and shape one's own life is intrinsic to being human. Freedom of choice is a radical freedom. Not even God seems to interfere with it. Some of the debates about the nature of grace focused on this very problem of free choice. Did God wait until the individual made a free choice to open up to God before bestowing grace? Or did God give what was called "Prevenient" grace to individuals, to assist them in making a choice for the good? While the former position underlined the importance of individual freedom in coming to God, the latter underlined the importance of God's assistance in all human affairs, to aid the natural stumbling of human reason and will. In any event, to ignore this radical human freedom is to violate the very humanity of the patient, a humanity given to him or her by God. Respect for patient autonomy is a professional obligation intrinsic to the moral nature of the clinical encounter.

On the other hand, the patient and his or her family also have an obligation in charity to respect the autonomy of the health professional or institution. The patient cannot, in the name of the absoluteness of autonomy, demand that the physician become the unquestioning instrument of the patient's will. The conscience of religious physicians or hospitals cannot be overridden any more than the physician can override the patient's conscience. Even if certain practices like abortion, sterilization, discontinuance of food and hydration or euthanasia are legally sanctioned, the patient cannot demand that the physician comply.

The problem of whether or not an institution can be said to "have" a conscience is an interesting one. Traditional views of ethics would rule out such a conception. In these traditional views, often held by physicians, only individuals have a conscience and can be held accountable for their actions. This is because ethics is taught by precept and example, instilling in the individual the virtuous practice of what is right and good. On this view, to speak of institutions having consciences is to waffle on the fundamental and individual choices that must be made. After all, institutions are merely collections of morally responsible individuals, and have no moral weight of their own.

Countering this view is one that sees institutions as social entities, capable on their own of being morally responsible for their acts and capable of being held publicly accountable for them. On this view, the trustees, directors, officers, etc. who make up the institution share collectively in the good and evil which the institution produces. It is far too easy to dismiss this view in a callous age in which many public institutions, like the public school system, a political party, or the CIA, continue to commit deeds that are held to be amoral. Yet an institution is not just a collection of individuals pulling in every direction. It is a cohesive and unified force for good or ill in society. Its administrators are held accountable for the actions of the entity itself, not just one or two individuals who happened to be employed therein. Hence, it is appropriate to speak of the collective "conscience" of an institution, if thereby, we mean the public commitment to values espoused by that hospital or healthcare organization.

The Christian, for example, could not accept the absolutization of patient autonomy and self-governance over life so forcefully promulgated by Judge

Compton in his concurring opinion in the Bouvia Case [5], or as might be argued by Engelhardt in his treatment on the origins of autonomy [10]. Judge Compton claimed that efforts by physicians and by a hospital to force Elizabeth Bouvia to accept fluids and nutrition (medically delivered food and water) were "unconscionable." In California, then, the conscience of the institution and of the doctors in this case were considered of much less interest than the interests of the individual patient, Bouvia, in attempting to commit suicide. In other states, the conscience of the institution and some caregivers is taken into account. When a decision is made to withdraw fluids and nutrition from an incompetent patient, as happened in Massachusetts in the Brophy case [7], the courts have suggested that the patient be moved to a facility which, and under the care of physicians who, would be willing to carry out the wishes of the patient. In this way, the conscience of the institution and some of the caregivers is preserved.

The clinical encounter, construed in terms of beneficence is, therefore, a balanced equation. It requires mutuality of respect for autonomy. It recognizes the moral agency of both the doctor and the patient. It makes of the clinical encounter a covenant of mutual trust and respect, not a contract for services in which the physician is a mere agent or the patient a mere consumer. Thus, the religious modification of autonomy in the clinical encounter includes the possibility of an "institutional conscience" precisely because institutions are integral participants in the covenant between the healer and the healed.

BENEFICENCE AND BEYOND

Christian notions of the virtue of charity, caritas, or agape derive their unique meaning from the life, teaching, and example of Jesus Christ. The central message of the Gospels is the announcement of the good news of a God of Love. He sent His only Son, who gave the ultimate demonstration of love when He suffered so that redemption might be made available to all humankind. Love of God and love of neighbor were Christ's daily exhortations to those who would follow him. Love is the touchstone to salvation, the only transforming power that can eradicate injustice from the world. Love is enjoined on all people because all are children of the same Father who loves us all.

Christian charity as epitomized in the Gospel went beyond the pagan notion of benevolence and beneficence, which could reach noble proportions on occasion.[4] Charity calls for love of all – enemy as well as friend, for the unjust as the just. Christian charity, as St. Paul so eloquently expounded upon it, called for a higher degree of self-effacement, to the doing of good out of the motive of love of God.[5] Its highest expressions call for sacrifice even to the loss of one's own life. Without other-directed love, the natural virtue of beneficence could become a mere tool of self-interest.

The Patristic tradition reinforced that gospel teaching by requiring charity due to the mutuality of obligation we owe each other because Christ is in every person [25]. It is Christ we love in our fellows and in ourselves. The more perfect Christian charity, the more it leads away from love of self and material goods and toward the dedication of self and material goods to the good of others. Christian love requires the kind of love that is free of utilitarian justification.

The differences between contemporary interpretations of beneficence and Christian interpretations of charity can be more concretely defined in health care ethics if we examine the spectrum of expressions possible for the principle of beneficence, philosophically construed. Thus, beneficence ranges from merely avoiding evil and/or harm to another, to doing good as long as it does not require inconvenience to the doer, to doing good at some cost to the doer, to doing good even at a great cost, and, finally, in the most heroic form, to sacrificing all one's good or one's life for another. Christian charity, or agapeistic beneficence, calls for wishing and doing good precisely under those circumstances in which it might be most difficult to justify doing so on rational grounds alone. Charity is, in some senses, "unreasonable" in that it violates philosophical standards of moderation [22]. It becomes reasonable only if we accept the fact of a revelation that counsels perfection and defines charity as entirely other-directed and for the specific reasons that the other is a child of God, loved by God, and, therefore, worthy of love by us. The other person, then, "becomes" God, to whom we can dedicate all our actions, in the spirit of Jesus' combining the commands of love of God and love of neighbor.

Given its roots in the Jewish and Christian traditions, it is not surprising that Western political and philosophical thought underscore this kind of respect for persons. On each view, each person is endowed with natural rights that no government, institution, or other person may rescind [23]. Moral philosophers, from as divergent moral philosophies as Kant and Mill, have argued the inviolability of the freedom of persons except when the freedom endangers others or the society as a whole.

This inviolability has also fed the Natural Law view that persons come endowed by their Creator with inalienable rights prior to the formation of communities and governments. The priority of natural rights over positive rights, those granted by communities and governments and their laws, is not so much a temporal priority, as if such rights were implanted in persons in the zygote stage, before the beings emerged from the womb and into the social and political community. Rather the natural rights have a metaphysical priority to those granted by society in positive law. That is to say, such rights trump any positive law that might impede them. This is a revolutionary idea that many countries still do not recognize. Such respect for persons, however, is a constant theme among religious leaders when addressing appalling poverty and the violation of human rights in countries around the world.

Clearly, both philosophical and theological reflection provide positive foundation for the central position of patient autonomy in all human relationships and especially the clinical relationship. What is different is the ontological pediment upon which respect for persons is based.

In Kant [13], Mill [15], W. D. Ross [21], and Beauchamp and Childress [2], for example, respect for persons or autonomy is a *prima facie* principle to be respected unless there is good reason for not doing so. There is, however, no derivation of this prima facie obligation. Instead there is more or less an intuitive justification based on coherence among thoughtful, reasonable, and well-intentioned people.

Theological ethics provides a firmer ontological ground for the centrality of patient autonomy in the existence of a common human nature, created by God, in the equality of all persons and their moral accountability before God, and in the examples of respect for persons based in charitable justice so evident in the life of Christ. Thus, on the Christian view, respect for persons has a deeper justification than simple human moral assertion, correct as that may be. Refer to our previous discussion about freedom. Philosophically based freedom is a characteristic of the individual with limits only of nonmalefiscence. One can do whatever one wants unless it might harm someone else. But theologically based freedom brings with it notes of responsibility. It is limited by duties of beneficence. One is free to do what one is obliged to do by nature and by grace for oneself and for others. Moral agency is implied by both philosophical and theological freedom, but the range of obligations is different.

The difference is not trivial, even if we recognize in the prima facie theorists and our own Declaration of Independence the echoes and remnants of the Christian teachings about the inviolability and worth of every human being. On this latter view, there is an additional obligation to respect autonomy that does not depend upon human determination. If respect for persons is argued only from an initial premise of acceptance by respectable people the question is begged. "Reasonable people" are defined as those who agree with the notion of autonomy which is what must be justified in the first place.

Moreover, the definition of "reasonable" is notoriously variable in a morally pluralistic society in which competing interests and contradictory moralities all piously assert their reasonableness. The same can be said of good will and intention. Moral sentiment and coherence theories might have had some substance in the 18th century when the Jewish-Christian tradition was still alive, though muted. Instincts could be appealed to precisely because there was still a residual common language and common expectation in the culture that was derived from the Judaeo-Christian heritage. "Coherence" in these theories, then, was implicitly, if not explicitly, based in the acceptance of the teachings of a theological ethic. Hard as the empiricists may have tried to escape and deny the roots of moral approbation in a religious view of the purpose of human life, those roots were still nutriment to their moral philosophies.

This is decidedly not the case today. The theological roots of human rights and of respect for persons are not only atrophic, but they are being energetically extirpated. The fact of moral pluralism has become a justification for the moral equality of all divergent opinions. "Reasonable people" might, at some point, agree that the dignity of persons and the respect for their autonomy are matters of societal convention which can be abrogated by mutual consent. Unfortunately, pathological societies (e.g., totalitarian states) do exist. Again, unfortunately, streaks of moral pathology run through otherwise morally decent societies (e.g., slavery, segregation, suppression of rights of women and children, and the treatment of aliens in wartime in our own society). These states of affairs once received widespread social approbation, but they were not, and could not be, morally defensible in a Christian ethic.

Only a foundation for human rights, for respect for persons and for autonomy, based in something more fundamental than a prima facie assertion can secure the fullness of respect that human beings owe one another. From the theological perspective, the dignity and equal worth of all humans is a divine command instantiated in every aspect of the life of Jesus [11]. On this view, the dignity of persons is assured no matter what states or political parties may deem "reasonable." This applies equally to those sad and indefensible violations of respect for persons perpetuated in the name of Christianity itself.

A theological perspective, one faithful to the Scriptural foundations of respect for human beings arising out of the Christian call to love our neighbors, is a more secure foundation for autonomy than philosophy, specifically in the clinical encounter. We make this claim on the basis of our argument that theological reflection can support a radical freedom metaphysically prior to that recognized by positive law. Patient autonomy, within the limits we have defined above, is an obligation to God as well as humanity.

If we add to this the special Old and New Testament concerns for the most vulnerable among us – the sick, poor, children, and the aged – we have a sounder, less fragile and more lasting assurance of patient autonomy than present medical ethical theory allows. On our view, autonomy, in addition to being a moral principle, becomes a moral claim of all humans upon all other humans, a claim we have a moral obligation to respect, especially in the healing relationship in which one party is so much more vulnerable than the other.

VULNERABILITY, TRUST, AND AUTONOMY

At the outset of this essay, we asserted that establishment of a morally equitable balance in the autonomy equation rests in large part on the restoration of trust, what we have called beneficence-in-trust [18]. It is important to begin with a recognition that most of the reasons for the growth of the autonomy movement are based in the realities of the clinical encounter and in actual and potential violations of the dignity of persons in the healing relationship.

First is the inevitable vulnerability of the sick persons, a phenomenon we have set out in detail in an earlier work [17]. Briefly put, the sick person is in a relationship in which he or she is dependent, anxious, in pain and distress, lacking knowledge of what is wrong, how serious it is, and of what can and should be done. The inequality in knowledge creates a predicament of inequality in which there is a real possibility of exploitation if the physician is not one of good character. Exploitation takes many forms – financial gain, enhancement of the doctor's prestige and power, imposition of his or her moral or social values, etc. The inherent vulnerability of the patient understandably generates a sense of fear and distrust and an accompanying desire to protect one's self against possible violation of trust by an unscrupulous doctor, nurse, hospital, or health care institution.

This ineradicable vulnerability is exacerbated by the moral pluralism of our society, by the fact that the doctor and patient are now so often strangers to each other. The probabilities of doctor and patient sharing the same values are increasingly remote. As our society becomes more and more polyglot ethically, linguistically, and socially, the fear of one's most deeply held beliefs being ignored increases exponentially.

To this background of moral diversity must be added the general distrust in the last three decades of all authority – of government, business, and institutions of all kinds, including universities. Even the idea of a profession is suspect. Professions are seen not in traditional terms as groups that might be expected, with some reliability, to suppress selfish self-interest, but as monopolies or, as George Bernard Shaw saw them, as conspiracies against the public.

Doctors, particularly, are counted as moral conservatives on matters like sexual freedom, abortion, euthanasia, recreational use of drugs, homosexuality, and "free-swinging" lifestyles in general. Doctors are perceived as prisoners of their high powered technologies and an outmoded ethical code which drives them to overtreat dying and terminally ill patients.

There is, unfortunately, empirical evidence that some physicians fit some of these caricatures. There are unconscionable fees, evidence of conspicuous consumption, arrogance, insensitivity to patients' needs or requests, and the boast that they (physicians) can get from the patient any decision they want by the way they get consent.

TRUST IN THE DOCTOR-PATIENT RELATIONSHIP TODAY

The move to patient autonomy arises in a combination of these causes and the generally legalistic climate of human relationships prevalent in our society. When we speak of balancing the autonomy equation or restoring trust to the healing relationship, we are denying neither the existence of these factors nor the necessity for both legal and moral curbs on maleficent paternalism. But, understanding the reasons for the move to autonomy and the necessity for legal and moral protection for the patient does not mean that the absolutization of autonomy – or what, at the present moment, really amounts to the abolition of trust – is morally defensible or wise. Let us turn to a consideration of what a theological perspective can contribute to a better balance in the equation of autonomy among and between the partners in the healing relationship and its relationship to trust.

The absolution of trust is what some reputable and thoughtful ethicists are proposing [6, 8, 24]. For them, trust is a dangerous concept. It imperils the patient's autonomy so seriously that it must be eliminated by measures that will give absolute supremacy to patient's wishes. They propose to accomplish this by reinterpreting the physician's role from medical authority to advisor and educator, supplanting the physician by a case manager, and converting the fiduciary relationship into a contractarian one. One of us has detailed why the elimination of trust is an empirical impossibility and why neither the contract model nor the displacement of the role of medical advisor to a non-physician is morally or logically tenable [19].

We agree with the critics of the trust model that it is difficult fully to fulfill the moral obligations of a fiduciary relationship, particularly one as complex as the healing relationship. The critics of trust do a service by sensitizing us to those complexities. But the difficulties do not excuse the physician from the moral obligations peculiar to healing. Trust cannot be eliminated, even if it were desirable to do so. Instead, we propose that those complexities impel the physician to an even more meticulous performance of her responsibilities, to act in the patient's interests while also preserving autonomy to the extent possible.

The current absolutization of autonomy gives testimony to the way mistrust has replaced trust in the physician-patient relationship. Some would go so far as to say that our insistence on the ineradicability of trust is merely a cover-up for the way we use trust to manipulate patients for our own purposes. A slightly more charitable accusation is that the physician's plea for trust is a subtle way to defeat the autonomy movement.

These accusations should not be dismissed lightly. There are physicians who do violate the trust relationship – consciously or unconsciously. There are also reputable physicians who insist that they can get any decision they want from the patient by the way they select and present the facts or manipulate the consent process. Some sadly boast about their abilities and use this fact to deny that genuine patient autonomy is an illusion.

There is no question about the possibility of abuse of trust since the patient is so highly vulnerable. But the reality of, and the opportunity for, abuse does not remove trust as a reality [20]. Attempts to eradicate it in the physician-patient encounter must inevitably fail. Such attempts end up either by displacing trust from the physician to some other person, who must be trusted, or trying to eliminate it by some rigid a priori contract – both of those attempts are self-defeating. In the end, both attempts move trust from the physician to some third entity, another person or a document. Trust will still be required nonetheless to fulfill the contractual obligations under these schema.

We would argue, instead, that the very fact that trust may be misused, together with its ineradicability, only multiplies the weight of the physician's moral obligation to remain faithful to trust. We could make this argument on philosophical grounds alone. It stems from the impersonality of the age. We live in a time when treatment is offered, not so much by our primary care physician who might be our friend in the community, but by strangers and specialists whom we must trust to care for us and respect our values. Personal values are difficult to preserve in the impersonal and depersonalized health-care environment of our day. Patients feel estranged from their physicians as a result.

This cuts both ways. Physicians, too, feel estranged from their patients. Caution about possibility of litigation, excessive demands on time and personal life, overwhelming burdens of debt after medical school, and professional burnout, are not only the result of social structures that permit and encourage dysfunctional behaviors between doctors and social relations, but also a direct result of impairments in the primary relationship with patients themselves [4]. Trust is required to bridge this isolation within the primary relationship of doctor and patient.

If the healing relationship has the special fiduciary characteristics we have ascribed to it earlier, how does a Christian perspective influence that relationship in such as way that fidelity to trust is enhanced? For the Christian, autonomy is framed in a context that accommodates the fact that our ineradicable dependence upon a Creator is the source of all we possess, are, and will be. For believers, life is a gift over which we have stewardship, but not absolute rights. Divine Providence enters life at every step and stage. Who we are, what we are to do and become, are things that we try to shape in accord with God's will for us. As we grow spiritually, our prayers move from petition to supplication, to intercession, to abandonment to the Will of God. The act of faith is an act of trust and acceptance, not of independence.

This does not mean that Christians place less value on the dignity of human beings or less on their moral accountability. Christians are no less protective of their moral right to participate in medical decision-making than non-Christians. Indeed, their commitment to a Source of moral value that transcends human existence only deepens their need for the autonomy which is necessary to express and to protect values such as the dignity of persons and moral accountability.

Faith gives a firmer foundation than mere philosophical assertion for respect for persons in their common origin as creatures of God. Faith also puts a limit on the hubris which the desire for autonomy so easily generates. Christians are expected to respect life, to preserve health, and to seek treatment when it is effective. They are not expected to cling to life when death is inevitable. Rather, they must confront their finitude and not expect technological salvation. Christians should enter the healing relationship, therefore, with a realistic awareness of the limitations as well as the power of medical interventions. Insistence on "doing everything" in the vain hope for medical miracles bespeaks a lack of trust in God. True miracles, i.e., interventions by God in the natural order of things, are not compatible with fruitless prolongation of life by all available means when death is unavoidable. Indeed, to look for a technological miracle is to lack faith in God's will and power to rescue us when He deems it appropriate and on His terms, not ours.

Clearly, too, the absolutization of autonomy to include active euthanasia or assistance in suicide seems to be inconsistent with a Christian faith commitment. To be sure, for Christians, dying, death, and suffering are not human experiences to be savored masochistically; they are events, nonetheless, grounded in original sin and atonement. They parallel events in the life of Jesus, himself. Effective treatment for the underlying disease and effective relief of pain are not to be eschewed. But we cannot use our autonomy and freedom to end our lives deliberately at the time of our own choosing. Nor can we ever accept the "good death" or aid in dying that asks for the doctor to kill us or aid us in suicide. There is a deeper meaning in our last days whose mystery we do not fully comprehend.

The whole character of the healing relationship is, thus, transformed when the patient and/or the physician is a believing Christian. The Christian patient is impelled to seek and use medical knowledge when it is effective, to act autonomously in accordance with conscience and Christian moral values, and to accept death, dying, and even suffering, when disease is beyond the means of medicine. As it says in Ecclesiasticus ben Sirach (38), we should honor the physician for we have need of him, but we must also never forget that the means of healing, the content of medical knowledge, and the physician, are, themselves, all manifestations of God's, not their own, healing power.

On the other hand, however much the healing acts of the physician may reflect Christ's own healing, the physician is not Christ. She is not to be accorded the authority of God, nor the act of abandonment we can show to God. Dependent as we may be upon the doctor's knowledge, we remain responsible moral agents, accountable for our own moral choices. We may freely yield up our moral claim to autonomy. But when we do so, we must require that those to whom we entrust it – our physicians and our surrogates – act within the spirit and intent of Christian ethics, i.e., within the context of a faith-inspired insight into being ill, being healed, dying, and suffering.

Non-Christian physicians who may reject the Christian teachings are required nonetheless, on purely naturalistic grounds of respect for autonomy, to respect the way a faith commitment reshapes the doctor-patient relationship. Not to do so, even in the name of beneficence, is to violate the very autonomy which secular ethicists so vigorously defend. The gulf between Christian and secular values in healing and the use of medical technology is widening daily. Doctors and patients with different religious belief systems must recognize those differences as soon as possible and sever their healing relationships before a crisis of moral values produces irresolvable conflict.

For the committed Christian, fidelity to trust is an obligation in charity. Its violation is tantamount to what used to be called "sins that cry to heaven for vengeance" – those sins against the most vulnerable amongst us, the widows, the orphans, children, etc. The Christian physician cannot violate trust, even for good ends like getting consent for effective treatment. The Christian physician must be as self-critical as possible and sensitive to the most subtle nuances of the trust relationship.

Fidelity to the requirements of a faith commitment would allow physicians and patients to preserve trust, in a relationship which cannot function without it. Restoring trust will facilitate healing in its broadest sense, avoid the minimalism of a legal contract relationship, and protect the patient's autonomy.

CONCLUSION

It is clear from the experiential, conceptual, and theological points of view that trust cannot be eliminated from human relationships, least of all relationships with professionals. Given this fact, an ethic based on mistrust and suspicion must, by the nature of human relationships, ultimately fail. To be sure, living wills, contracts if one wants them, durable powers of attorney, or appointment of a patient advocate or health care manager can diminish some of the vulnerability of trust relationships. In the end, however, all of these arrangements attempt to displace trust to some degree from the physician or professional and locate it elsewhere – but trust still remains as guarantor of the deed itself. This guarantee, we suggest, stems from the very nature of human interactions, the vulnerability of those engaged in the healing encounter, and the commitments necessary from a faith perspective to each human being as created and prized by God.

The theological perspective on the clinical encounter, then, contributes to a more profound understanding of the meaning of autonomy and trust in the doctor-patient relationship. First, autonomy is modified to include one's relationships at an ontological level. Second, trust is rooted in that ontological relationship, prior to explicit and contractual arrangements that might enhance it (not replace it). Third, understood from this deeper perspective, any violation of autonomy would harm healing. Fourth, a relation of one's life with a higher power re-interprets autonomy as a stewardship of a gift rather than as pure self-determination. As we have taken pains to show, however, this sense of "autonomy" does not diminish the obligation of self-actualization. Rather it enhances it in the context of human nature and the healing relationship.

Loyola University of Chicago Medical Center
Maywood, IL
U.S.A.
and
Georgetown Institute for the Advanced Study of Ethics
Washington, DC
U.S.A.

NOTES

[1] This universalism was especially evident in Zoroastrianism and Judaism. Among Jewish prophets, Second Isaiah is generally regarded as the first to formulate a universalism based upon a theology of creation.

[2] The Greek relationship was described as one of friendship, the Christian as one of love, and the contemporary as one of comradeship. All of these presuppose respect for persons.

[3] For a secular view of that antagonism, see [2].

[4] For an example of some lofty Stoic sentiments, see Scribonius Largus' Introduction to *Compositiones* (cited in [9], pp. 336–344). Also cited is Libanius' speech to young physicians exhorting them to cultivate "love of man," to "share the pain" of the patient ([9], p. 345).

[5] For an exposition of Paul's notion of Christian justice and love, see [12], pp. 282–288.

BIBLIOGRAPHY

1. Abbott, W., (ed.): 1962, *The Documents of Vatican II*, Geoffrey Chapman, London.
2. Beauchamp T. L., and Childress, J. A.: 1981, *Principles of Biomedical Ethics*, Oxford University Press, New York.
3. Beauchamp, T. L., and McCullough, L. A.: 1984, *Medical Ethics: The Moral Responsibilities of Physicians*, Prentice-Hall Publishers, Englewood Cliffs, NJ.
4. Bergsma, J., with Thomasma, D. C.: 1983, *Health Care: Its Psychosocial Dimension*, Duquesne University Press, Pittsburgh, PA.
5. *In Re: Bouvia v. Superior Court of the State of California for the County of Los Angeles*, 225 Cal. Rep. 297, (Ct. App. 1986), review denied (Cal. June 5, 1986).
6. Brock, D. W.: 1991, 'Facts and Values in the Physician-Patient Relationship', in E. D. Pellegrino, R. M. Veatch, and J. P. Langan (eds.), *Ethics, Trust, and the Professions, Philosophical and Cultural Aspects*, Georgetown University Press, Washington, D.C., pp. 113–138.
7. *In Re: Brophy v. New England Sinai Hospital, Inc.*, 398 Mass. 417, 497 N.E. 2d 626 (1986).
8. Buchanan, A.: 1991, 'The Physician's Knowledge and the Patient's Best Interest', in E. D. Pellegrino, R. M. Veatch, and J. P. Langan (eds.), *Ethics, Trust, and the Professions, Philosophical and Cultural Aspects*, Georgetown University Press, Washington, DC, pp. 93–112.
9. Edelstein, L.: 1967, *Ancient Medicine, Selected Papers of Ludwig Edelstein*, Owsei Temkin, and L. C. Temkin (eds.), Johns Hopkins University Press, Baltimore, MD.
10. Engelhardt, H. T. Jr.: 1986, *The Foundations of Bioethics*, Oxford University Press, New York.
11. Graber, G. C.: 1975 'In Defense of a Divine Command Theory of Ethics', *Journal of the American Academy of Religion* XLIII, 62–69.
12. Haughey, J.: 1977, 'Jesus as the Justice of God', in J. Haughey (ed.), *The Faith that Does Justice, Examining the Christian Sources of Social Change*, Paulist Press, New York, pp. 282–288.
13. Kant, I.: 1965, *Metaphysical Foundations of Morals*, J.C. Freidrich (ed.), Modern Library, New York.
14. Lain-Engralgo, P.: 1969, *Doctor and Patient*, McGraw-Hill Book Company, Inc., New York.
15. Mill, J.S.: 1956, *On Liberty*, C. V. Shields (ed.), Bobs-Merrill, Indianapolis.
16. O'Rourke, K.: 1989, 'Should Nutrition and Hydration be Provided to Permanently Unconscious and Other Mentally Disabled Persons?', *Issues in Law and Medicine* 5(2), 181–196.

17. Pellegrino, E. D., and Thomasma D. C.: 1981, *A Philosophical Basis of Medical Practice*, Oxford University Press, New York.
18. Pellegrino, E. D., and Thomasma D. C.: 1988, *For the Patient's Good, The Restoration of Beneficence in Health Care*, Oxford University Press, New York, NY.
19. Pellegrino, E. D.: 1989, 'Character, Virtue, and Self-Interest in the Ethics of the Professions', *The Journal of Contemporary Health Law and Policy* 5, 53–73.
20. Pellegrino, E.D.: 1991, 'Trust and Distrust in Professional Ethics', in E. D. Pellegrino, R. M. Veatch, and J. P. Langan (eds.), *Ethics, Trust, and the Professions, Philosophical and Cultural Aspects*, Georgetown University Press, Washington, DC, pp. 69–92.
21. Ross, W. D.: 1988, *The Right and the Good*, Hackett Publishing Company, Indianapolis, IN.
22. Shelp, E. (ed.): 1985, *Virtue and Medicine: Exploration in the Character of Medicine*, D. Reidel Publishing Company, Dordrecht/Boston, essay
23. Thomasma, D. C.: 1991, *Human Life in the Balance*, Westminster Press, Louisville, KY.
24. Veatch, R. M.: 1991, 'Is Trust of Professionals a Coherent Concept?', in E. D. Pellegrino, R. M. Veatch, and J. P. Langan (eds.), *Ethics, Trust, and the Professions: Philosophical and Cultural Aspects*, Georgetown University Press, Washington, DC, pp. 159–173.
25. Walsh, W. J., and Langan, J. P.: 1977, 'Patristic Social Consciousness', in J. Haughey (ed.), *The Faith that Does Justice, Examining the Christian Sources of Social Change*, Paulist Press, New York.

DANIEL P. SULMASY

EXOUSIA: HEALING WITH AUTHORITY IN
THE CHRISTIAN TRADITION

The contemporary Western world holds tenaciously to a demand for individual liberty which stands radically opposed to an ever increasing need for individuals to be dependent upon the expertise of others. Ironically, the banner of autonomy has been raised high at a moment in history characterized by profound interdependence in a complex, specialized, technological culture. Perhaps this conflict between the demand for independence and the demand for dependence is nowhere more readily apparent than it is in medicine.

It is easy, then, to see why the nature of authority should be such a thorny problem for contemporary society, and particularly for the practice of medicine. The many contemporary meanings of the word "authority" are perhaps reflective of the depth of the doubt about what authority actually is. This confusion is exemplified by the common observation that patients today often go to great lengths to get authoritative opinions regarding their various conditions, only to feel victimized because the very physicians who have rendered such opinions have treated them authoritatively.

Thus, the role of authority in the relationship between doctors and patients is usually understood as a struggle for power between doctors and patients. Power is generally understood as force of knowledge or force of will, and contemporary ethical arguments about authority in the doctor-patient relationship can usually be characterized as advocating either more or less power for either doctors or patients. But the central thesis of this essay is that, from a theological perspective, the whole basis of these arguments is wrongly conceived. The Judeo-Christian notion of authority is best expressed by the Greek word for authority, *exousia*. The *exousia* to heal which Jesus gave his disciples (Mt. 10:1, Mk. 3:15; Lk. 9:1), has nothing to do with a struggle over knowledge and will between doctors and patients. To heal with *exousia* is to heal with an understanding that the only legitimate power (*dynamis*) expressed in the doctor-patient relationship is the *dynamis* of healing itself, and that this *dynamis* has a source which transcends and subsumes that relationship. A medical practice informed by *exousia* might negotiate a new course, avoiding the pitfalls of both unconstrained patient autonomy and physician paternalism. The perspective of *exousia* provides an alternative vision for medicine at a time when the fragmentation of an individualistic medical marketplace and the bureaucratic dehumanization of an unbridled medical technocracy threaten the integrity of the entire medical enterprise.

G.P. McKenny and J.R. Sande (eds.), Theological Analyses of the Clinical Encounter, 85–107.
© 1994 Kluwer Academic Publishers. Printed in the Netherlands.

THE MANY MEANINGS OF AUTHORITY

The meaning of authority is anything but clear. Kierkegaard's complaint about the "confusion involved in the fact that the concept of authority has been entirely forgotten in our confused age" [18] remains valid even today. To help set the many meanings of authority into a framework from which analysis can proceed, I will place these meanings under three basic headings. I will designate an appropriate preposition or article for each of these three uses of the word. Finally, I will define the three senses in which the adjectival form, "authoritative," is most closely associated with each meaning of the noun, "authority."

1. *Authority as Control*

The most typical meaning of authority refers to force of will or the ability of one person to control another's thoughts, words, or deeds. It can refer either to the controlling power that a person actually possesses (e.g. – she is in a position of authority), or it can function as a noun designating the person who is in control, often in the plural (e.g. – she is wanted by the authorities). The typical preposition associated with this usage is "in." To be *in* authority is to have control. Using the word in this sense in a medical context one might say, "Doctors have too much authority." When the adjectival form, "authoritative," is used, it typically refers to an abuse of power or control (e.g. – that surgeon behaves authoritatively).

2. *Authority as Expertise*

The word "authority" is also frequently used when referring to knowledge, skills, precedents, and conclusive statements. It is especially used as a noun to refer to one who has such knowledge or skill. The typical article associated with this usage is "an." To be *an* authority is to possess knowledge and skills superior to others, often rendering the others dependent upon the authority for access to some good or service. Using the word in this sense in a medical context one might say, "Dr. Jones is an authority on ocular melanomas." When the adjective "authoritative" is used in this sense, it means that the opinion or answer is conclusive.

3. *Authority as Warrant*

A less common but by no means archaic use of the word refers either to the freedom granted by one who is in control, or to actions carried out with conviction. The preposition linked to this usage is "with." To act *with* authority is to act in the freedom granted one by someone else or to act with an apparent sense of legitimacy or conviction. In a medical context, using the word this way sounds unfamiliar. But one might say, for example, that "Dr. Smith practices with authority." When the adjective "authoritative" is used in this sense, its meaning can sometimes be approximated by the adjective "legitimate," sometimes by the adjective "genuine," and sometimes by both.

These three clusters seem to capture the families of meaning which come under the broad term "authority." They overlap, of course, and may exclude some marginal meanings of the term, but this classification ought to be sufficient for the purposes of this essay.

AUTHORITY, HOBBES, LOCKE, AND MEDICINE

The use of the word "authority" in political philosophy depends heavily upon the English philosophers Thomas Hobbes and John Locke. This approach has been uncritically accepted by many as the primary means of understanding the role of authority in the doctor-patient relationship, particularly with respect to the concept of informed consent ([6], pp. 76–77; [8], pp. 44–47, 267–268; [10], pp. 13–14, 174–175, 369–373; [44], pp. 190–213). These authors use the word primarily in the first sense (authority as control). The word is also sometimes used in the second sense (authority as expertise), but generally with the assumption that expertise implies control (i.e., – to the extent that the doctor is *an* authority, the doctor is *in* authority). To understand the roots of this conception of authority, one must go to the sources.

For Hobbes, the questions surrounding authority begin with his convictions about human nature. "I put for a general inclination of all mankind a perpetual and restless desire of power after power, that ceaseth only in death," [*Leviathan*, Ch. XI]. Human beings, by nature, seek power, by which Hobbes means control. Liberty, for example, is simply the absence of external control [*Leviathan*, Ch. XIV]. Hobbes' convictions about ownership, coupled with his convictions about an innate human desire to control others and to be free from their control, defines what he means by authority. "He that owneth his words and actions is the AUTHOR: in which case the actor acteth by authority. . . . And as the right of possession is called dominion, so the right of doing any action is called AUTHORITY," [*Leviathan*, Ch. XVI]. Authority, then, is defined negatively: the absence of external control in the disposition of the actions and words one possesses. Finally, Hobbes is convinced that one will give up one's claims to authority only for greater gain, either by individual contract [*Leviathan*, Ch. XVI], or for the sake of self-preservation through participation in the commonwealth [*Leviathan*, Ch. XVIII]. Thus, the Hobbesian conception of authority essentially requires only two things: "effectively uncontested power and the right to rule," [19].

Locke's concept of authority is likewise connected to his concept of political power, which is described in terms of control. "It is impossible that the rulers now on earth should make any benefit, or derive any the least shadow of authority from that which is held to be the fountain of all power, 'Adam's private dominion and paternal jurisdiction'. . . ." [*Second Treatise on*

Government, Ch. I]. Closely connected is Locke's concept of negative rights, which limit "the extent of the legislative power." Individuals are not to be interfered with, and others are given "power to make laws but by their own consent and by the authority received from them" [*Second Treatise on Government*, Ch. XI].

The key to understanding authority in contemporary political philosophy is to be ever aware of its Hobbesian/Lockean roots. Power, considered as the ability to control and be free from the control of others, is the implicit assumption which dominates contemporary discussions of authority. Friedman, for instance, suggests that whether one considers authority as the ability to rule or influence (*in* authority), or the ability to inspire belief (*an* authority), it is always control which is at issue. He argues that one must surrender control by surrendering private judgement either in obeying a command or in accepting a premise on authority [14]. Similarly, Raz [36] admits that the contemporary notion of authority comes from a coercive concept of law. However, he suggests that authority be thought of primarily as a moral right to impose a duty, and only secondarily as a right to coerce others into compliance with these duties. But his bottom line is coercion, and the moral right to impose a duty must still be understood as control. Therefore, in contemporary political philosophy, authority is inevitably seen in conflict with autonomy [46].

Both the impetus for the dramatic new role of patient autonomy in medical decision making and the recent evolution of the doctrine of informed consent have depended upon a Hobbesian/Lockean conception of authority in the doctor-patient relationship. Flathman [12], for instance, paints a Hobbesian picture of doctors and patients. He sees two basic models of authority in the relationship:

1. a consensual model in which the physician is authorized to practice only so long as the physician's actions are congruent with the consensually agreed upon values of the community, and
2. a constrained conflict model in which there is no general consensus, and so physicians are placed *in* authority inasmuch as patients will do what physicians say despite their disagreement.

Flathman feels that the latter is more realistic. Patients reluctantly accept dependence upon experts because the price of not doing so is to give up the services the experts provide. Power dominates Flathman's discussion. Knowledge is power. Power is control. Therefore, as experts, physicians exert control through the power of their knowledge. And thus, life in the waiting room is inevitably solitary, poor, nasty, brutish, and short [cf. *Leviathan* Ch. XIII].

Veatch [43] is also influenced by Locke and Hobbes. He details how physicians do, in fact, act coercively, by controlling access to hospitals and drugs, committing suicidal or psychiatric patients to involuntary admissions, or administering required immunizations. Veatch accepts, however reluctantly, the necessity of giving such control to physicians. Physicians have the expertise to protect the healthy and the sane from the contagious and the psychotic. But while Veatch denies that there are any "value-free facts," he also claims to be able to distinguish facts from values in medical decision making. He insists that the authority of physicians be limited wherever possible to the technical arena, which is more factual than evaluative. He argues that the physician should be *in* authority only to the extent that the physician is *an* authority. In effect, Veatch argues that patients own not only their bodies, but also the evaluative ideas they have about illness and treatment. Since medical decision making inevitably involves not only the patient's body but also all that the patient values, all medical authority properly belongs to the patient. Even though he thinks it is epistemologically impossible for a real physician to do so, Veatch's ideal physician would dispense "valueless," objective information about the body to patients, who must, regrettably, depend upon physicians for this information. Patients would then be independent in their decision-making. What is at stake in Veatch's theory is, of course, control. The authority of physicians is based on their technical expertise which defines, for Veatch, the moral limit of their control over patients. He writes, "No one in his right mind would conclude that those who are custodians of a particular value [knowledge of the body] should bear the responsibility for resolving disputes over the relation of that value with other values leading to one's integrated wholeness" [43]. Veatch worries that physicians, like Locke's princes, may overstep their prerogatives. Veatch's solution seems to be equally Lockean. He seems to urge patients, as Locke once urged the prince's subjects, "to get prerogative determined in those points wherein they found disadvantage from it" [*Second Treatise on Government*, Ch. XIV]. That is, Veatch seems to argue that the controlling authority of physicians can be to the disadvantage of patients and hurt them. Therefore, patients should seize control of those liberties traditionally given to doctors which have led to problems for patients. In declaring limitations on the doctor's latitude, Veatch could easily quote Locke and insist that no physician should complain about such a program of transferring control from physicians to patients, "because in so doing they [the patients] have not pulled anything from the prince [physician] that of right belonged to him" [*Second Treatise on Government*, Ch. XIV].[1]

THE INADEQUACIES OF THE CONTROL MODEL

There are several underlying assumptions in the control model of authority which must be critically examined. These assumptions are foundational. They are so deeply embedded in the ethical theories which flow from them that they often escape attention. But an exposition of these assumptions seems necessary in order to explain some of the difficulties one faces in considering the concept of authority in contemporary medicine, and to look for fruitful alternatives.

HUMAN NATURE

First, these Hobbesian and Lockean theories of authority make implicit but striking assumptions human nature. These assumptions about human nature are certainly not value-free. Notwithstanding Veatch's insistence that physicians be strictly limited to evaluative judgements about the body as such, the theories of both Flathman and Veatch begin with sweeping evaluative assumptions about the nature of the actors in the doctor-patient relationship. First, these theories assume that human beings constitutively seek personal liberty and control over others. Therefore no one, whether a doctor or a patient or a Native American Chief, is worthy of trust. Second, these theories take the voluntary contract forged between equals to be the paradigmatic human ethical interaction. Therefore, as Hauerwas [16] has observed, these theories presuppose that "all relations that are less than fully 'voluntary' [are] morally suspect."

But such assumptions are largely untrue, particularly in the medical context. First, while acknowledging the reality of sin, it must be argued why one should accept the Hobbesian notion that the primary human drive is to control others and be free of their control. Christian belief, for example, suggests that the primary drive is to love and to be loved. In fact, most persons *can* name other persons that they can trust, and many would place their physicians on their list. Only the most distraught and disheartened say, in their alarm, that no one can be trusted [Ps. 115:2]. Those who cannot count their physicians among the trustworthy generally want another physician, because they understand the critical importance of trust in the doctor-patient relationship.

Second, the most paradigmatic human interactions are not voluntary contract interactions between equals, but involuntary relationships between unequals [4]. The most important human relationships are the ones over which people have no control. No amount of innovation in reproductive technology will ever allow people to choose their own biological parents. Each person enters this world helpless, completely dependent upon others. People have no power to declare themselves immortal or free of the possibility of disease. These conditions are out of human control. And it is precisely in the midst of this absence of control and in relation to the state of dependency which illness engenders that the ministrations of medicine are meted out.

Veatch notes the inadequacies of "raw contracting" as the moral basis of the relationship between doctors and patients, but he unfortunately merely replaces the notion of raw contracting with a boiled down form of contracting [43]. In so doing, he continues to cling to the notion that control is the basis of the doctor-patient relationship. But this view is contradicted by a reality which cannot be otherwise. The doctor-patient relationship is predicated firmly on the fact of illness, which entails the loss of control.

Finally, the increasing interdependence which constitutes our contemporary social relationships, particularly in the medical arena, ought to provide a clue that human beings are not inherently atomistic, but inherently social and interdependent. As Aquinas put it, "man has a natural inclination to know the truth about God and to live in society" [*Summa Theologiae* I,II, q. 94, art. 2.c]. It would seem obvious this is neither Veatch nor Flathman's view of human nature.

Human beings are flawed, of course, and often fail to live out their potential. But a judgement that human beings are naturally selfish and that any behavior which appears to be goodness is really self-interest merely begs the question. Nor is this a matter which can be settled by experience. Experience teaches us only the following: some doctors are mostly good, and some doctors are mostly bad. It will remain an axiomatic choice, a faith assumption, to decide whether the fragile vessel of the physician is a glass half empty or a glass half full. The assumption of the Roman Catholic tradition of Christianity is optimistic: grace can build upon the reasonableness of human nature. Other Christian traditions are not so optimistic about the state of human beings outside of grace, but are at least optimistic to the extent that they believe in the power of grace to fill the fragile human vessel. In contrast to the Hobbesian assumption, Christian belief points to an open possibility that human beings can be better than they are now. Neither physicians nor anyone else will ever be better unless this possibility can be assumed.

AN HMO WITH ONLY ONE MEMBER?

The second problem with the control model is that it assumes an intersubjectivist morality for medicine. It assumes that all moral truth in medicine resides in the subjectivity of the autonomous individual. But since individual patients need other individuals called doctors when they seek healing, this poses a problem. More than one subjectivity is involved once a person enters a human relationship. How is one to settle differences if neither has a greater claim to be in the right? The only possible solution under the assumptions of the control model is to construct an *intersubjective* morality, either by contract or consensus. But since such intersubjectivity is never quite objectivity ([22], p. 22), the project is doomed to fail.

Veatch [43] and Engelhardt [8, 9] appear to argue along the following lines. They begin with the assumption that each individual is his or her own moral authority. As Flathman argues [12], when there is complete consensus on what the good is for medicine, there is no need for any external authority in medicine. But Veatch and Engelhardt both agree that such an intersubjective consensus does not exist. Therefore, they conclude that patients and doctors whose views overlap ought to seek each other out, forming voluntary communities of intersubjective agreement in medical morals in which authority can function.

The problems with this view are significant. Taken to its logical conclusion, the theory implies that each person ought to become his or her own personal Health Maintenance Organization (HMO). If there truly is no source of authority (conceived of as control) other than oneself, and conflict is inevitable because human nature implies a need to control others and to be free from their control, then each individual would ideally be his or her own personal health care system, in complete control of his or her own care and free from the control of doctors. Ideally, one supposes, all medical information about one's own body could be processed and analyzed by a computer which would be programmed to provide the treatment one selected from a range of options.

But such a view is far from reality and not ideal for anyone. Medicine is an intrinsically interpersonal enterprise. A purely rational computer medicine could never truly *care* for patients. Care requires persons. Yet Veatch's ideal of value free information given to completely autonomous patients who are free to decide what to do with that information in light of their own subjective values can only be realized by a machine. No one of right mind would want to be cared for by a health care system which merely objectively provided information about the body and paid no attention to the value of the whole person. There is even emerging evidence that the doctor-patient relationship is itself part of the therapeutic effect [38]. In addition, medicine seems to require (at least in those important cases where people seek out doctors) the presence of another. Self-diagnosis and self-treatment are always dangerous, even for experts. Finally, it seems that medicine is not just an interpersonal interaction between two individuals. It is an inherently communal enterprise. Without a prior commitment of professionals to share knowledge with each other, a system of one-person HMOs, if ever started, would soon grow into an absurd system of isolated, proprietary medical data banks, limited by the narrow experiences of individuals who functioned as their own doctors. The Hobbesian would then face a dilemma. Since medical knowledge is control, and control is what the Hobbesian desires, to share his medical knowledge would be to relinquish precious control over others. In addition, sharing implies accepting information from others, and to do so would be to acknowledge their control over his life, and this too would be unacceptable to the Hobbesian. On the other hand, not to share might lead like-minded persons to be equally stingy, and then he would risk dying from a curable sickness that he would not have the knowledge to treat. Thus he would neither be able to share nor not share. The Hobbesian view cannot be sustained in the limit.

Now a Hobbesian might not concede that there is a problem with his assumptions. Painfully, the Hobbesian might say, the above ideal of private, value-free computer medicine is simply not possible, even though it really *is* what everyone would want. Therefore a Hobbesian patient would reluctantly compromise for the sake of personal interests and accept dependence on medical professionals, but only to the least degree compatible with the patient's interests in pleasure, health, and longevity.

But the counter-argument here is standard. If the real justification for accepting the control of others were the maximization of one's own best interests as one defines them, then the only consistent position would be to lie about one's acceptance of the controlling influence of others in order to gain the benefits offered by contract, but then to do as one pleases in order to escape from the control enjoined by the contract. For example, suppose that a Hobbesian smoker were to join an HMO which forbade smoking in order to eliminate the costs of caring for smoking-related diseases, thereby saving money for everyone in the HMO. The Hobbesian who loved to smoke might promise not to smoke in order to join this HMO and save money. But his actions would be most consistent with the underlying justification of the Hobbesian theory (i.e. – self-interest), if he smoked whenever he could do so without getting caught. It is easy to see that once this process became generalized, the very basis for the compromise reached by the social contract would be destroyed. Thus the Hobbesian HMO, whether with one member or many, results in a *reductio ad absurdum*.

MEDICAL MONASTICISM

As Finnis points out, groups can coordinate action to a common purpose or goal either through unanimity or authority ([11], pp. 231–233). But if there is no unanimity, and if the Hobbesian view is absurd, where can one look for a theory of authority in medicine?

MacIntyre notes that when groups cannot achieve what they must by acting as individuals, practices spring up to achieve those goals. Practices are not forms of political or organizational power, but organized, rule-governed enterprises requiring judgements about how to best understand particular cases or reformulate the rules in the light of particular cases [24]. Medicine, of course, is a practice. And practices are inherently prescriptive. The doctor is said to *prescribe* therapy. To make a prescriptive statement such as, "x ought to be done" is to make "a claim which by the very use of the words implies a greater authority behind it than the expression of feelings or choices" ([23], pp. 51–52). This is true no matter how much one may claim to "own" these feelings or choices.

MacIntyre [24] realistically surveys the contemporary West, in constant rebellion against all forms of tradition and authority, deeply divided and unable to form a consensus about anything, and wonders only why it has taken so long for society to come to the impasse now faced by medicine: the demand for absolute autonomy in the face of its increasing impossibility. In the dissolution of the culture he sees only a profession which has become, "not quite a craftsman's guild, not quite a trade union of skilled workers, not quite anything." He despairs of the possibility of ever achieving enough agreement on the nature and goals of medicine to ever have a true professional practice again. His only positive solution is the possibility of achieving small communities of patients and doctors with a common vision – a vision in which the Western world would be dotted with a series of HMOs operating as medical monasteries in these new Dark Ages. The West only awaits a "new St. Benedict" ([25], p. 263) who will be the founder of these medical communities.

But this view is ultimately also unsatisfactory. MacIntyre is right in calling medicine a practice and right that the institution of medicine is currently threatened by contemporary views of authority. But the practice of medicine still retains enough internal coherence to remain a unified practice. The bodies of atheists and of Christians remain, after all, fundamentally the same. While their ultimate moral views remain radically different, it is hard to see, in the end, how the proposals of Engelhardt ([8], pp. 366–369), Veatch [43], and MacIntyre [24], really differ, except that they vary in the extent to which each thinks that a series of distinct medico-moral communities is a goal to strive for or a state of affairs for which one might reluctantly settle.

AUTHORITY, SOCIOLOGY, AND MEDICINE

Political philosophy is not the only contemporary discipline with important views about authority in the doctor-patient relationship. Sociologists have a view of this relationship as well. The sociological view, however, shares important similarities with the view of political philosophy. The sociological understanding of authority is largely derived from the seminal work of Max Weber, who defined authority as the power to issue commands that will be obeyed ([45], p. 152). As in the work of Hobbes, conflict and control are the essential features for Weber. He did distinguish *Macht*, "the probability that one actor . . . will be in a position to carry out his own will despite resistance," from *Herrschaft*, "the probability that a command with a specific content will be obeyed by a given group of persons." But *Herrschaft* is still conceived of as one will controlling another will. It is simply a less overtly violent imposition of one will upon another. Weber distinguishes three types of justification offered for authority other than simple *Macht*: rational grounds, traditional grounds, and charismatic grounds ([45], pp. 324–352). But it seems, as Hauerwas has noted, that even Weber's typological tryptic "fails to clarify what it means to acknowledge an authority as legitimate" [16].

Talcott Parsons has addressed the issue of authority in the doctor-patient relationship forthrightly ([30], pp. 441–442, 464–465). While not referring to Weber directly in this regard, it is clear that Parsons also assumes a model of authority based on relationships of power. Parsons suggests a "social control" model based on the advantages to society of giving physicians control over individual patients. This theory results from Parsons' empirical observations. Yet his conclusions are undeniably *interpretations* of his empirical observations and cannot simply be unquestioningly accepted as factual. And even if it is the case that the interpretation of Parsons is true (namely, that doctors really do act as authority figures exerting social control over the ill), it cannot therefore be concluded that this is the way things *ought* to be. This would represent a genuine example of the "Naturalistic Fallacy"; a true violation of the fact/value distinction. The fact/value distinction requires that moral claims not be justified solely on the basis of factual claims ([5], pp. 336–379). On the strength of this principle, even if Parsons were correct in the judgement that physicians used authority to control patients, this would not imply that an interpretation of medical authority based on social control theory is morally correct.

These interpretations of authority as control, based on either sociological theories or the theories of political philosophy or both, are pervasive in the literature of medical ethics. Countless discussions of the conflict between autonomy and beneficence have essentially been based upon this interpretation. It has almost begun to seem as if the fundamental task for medical ethics is to find the proper balance of authority in the power relationship between physicians and patients, with beneficence interpreted to mean authority for the doctors, and autonomy interpreted to mean authority for the patient. For instance, in their book on informed consent, Faden and Beauchamp acknowledge that "the issue of proper authority for decisionmaking is an implicit theme throughout this volume. In health care, professionals and patients alike see the authority for one decision as properly the professional's and authority for other decisions as properly the patient's" ([10], pp. 13–14). Empirical researchers have used this schema in part, perhaps, because it is amenable to quantification on scales generated by survey instruments and seems to capture at least some of the reality of the interactions between doctors and patients. I myself have fallen into this trap [41], but I am now convinced that the model so constrains the rich reality of the doctor-patient relationship that it is inadequate. A solution is not to be sought by accepting the basic correctness of the model and merely suggesting a shift from "unquestioning acceptance of physician authority, as embodied in the Parsonian model" to a "more egalitarian bargaining" state [20]. The problem lies with the Hobbesian assumptions of the sociological model itself.

EXOUSIA, DYNAMIS, AND HEALING

When Jesus sent his disciples out into the world, he gave them "power and authority to overcome all demons and to cure diseases. He sent them forth to proclaim the reign of God and heal the afflicted" [Lk, 9:1–2; cf. Mt. 10:1 and Mk. 3:15]. In this passage, it is important to note that Luke attributes to Jesus a distinction between the power that heals (*dynamis* in the Greek) from the authority (*exousia*) to heal. This is a distinction which is made with remarkable consistency throughout the writings of the New Testament [1, 13, 15, 27]. In making this distinction, it would seem that Scripture is suggesting that neither force of will nor the power of expertise is at issue in a discussion of the authority to heal. This is a perspective which is remarkably different from any account of authority and healing based on Hobbesian/Lockean political philosophy or sociology.

In relation to healing, *dynamis* is the power of healing itself. It was *dynamis* that Jesus felt go out from him when the woman with the hemorrhage touched his cloak and was cured [Mk. 5:30]. *Dynamis* is the pure power to heal. *Dynamis* is power *for*, not power *over* others. In a neo-Platonic sense, *dynamis* is self-diffusive. It has nothing to do with force of will. *Dynamis* goes out from Jesus without his willing it.

Dynamis is also used to characterize expertise. Thus, Simon Magus, the magician, was said to have the *dynamis* to heal [Acts 8:9–25]. But *dynamis* is clearly distinguished from *exousia*. When Simon Magus eventually came to faith, he realized the insufficiency of mere *dynamis*. He also wanted the authority (*exousia*) to impose hands. But the apostles would not grant him that. The very fact that he wanted to buy *exousia* was an indication to them that he was unworthy. And when the Pharisees wanted to know how Jesus had the *dynamis* to forgive sin [Mk. 2:1–12], he avoided the word *dynamis* in his reply. He said, instead, that he had the *exousia* to both heal and forgive sin. He proceeded to demonstrate both.

Exousia presumes *dynamis*, but not vice-versa. In Greek usage *exousia* was an illusion if not backed by real *dynamis* [13]. *Exousia* meant "the warrant or the right to do something" [27]. Thus, *exousia* is really closest in meaning to the third definition of authority set forth at the beginning of this essay. *Exousia* denoted an inner sense, and even a "moral power" in Stoic thought [13]. In the New Testament, *exousia* refers to the rule of God in nature and in the spiritual world, and especially the freedom which is given to Jesus and which he gives to the apostles [13]. While *exousia* is exercised with respect to sickness, the elements, and demons, Jesus specifically rejects any political application of *exousia* [1]. His kingdom, as he tells Pilate, is not of this world [Jn. 18:36]. *Exousia* is intrinsically related to the *Logos*. Nothing takes place apart from the *exousia* of Jesus. It is the freedom given to the community which orients itself to the Word made flesh. Hence, it can never be used arbitrarily [13].

New Testament *exousia* cannot be bestowed or produced. It emerges in practice. It is not mere Weberian charismatic authority, which can be used for either good or for evil. *Exousia* rests upon a practical and convincing insight into the Good, the True, and the Beautiful. It springs forth out of tradition. It becomes manifest upon recognition by the community ([15], p. 17). Hence, the magician who already has the *dynamis* to heal cannot buy *exousia* [Acts 8:9–25]. God rebukes those who misuse power as raw *dynamis* and deny the *exousia* of God [cf. Is. 5:8–9]. *Exousia* comes with experience, and is characterized by wisdom, equanimity, talent, charisma, and selflessness ([15], p. 17). William Osler himself could scarcely have done a better job of describing the virtues of a good physician. A physician might have the *dynamis* of actually being *an* authority and even, to some appropriately limited extent, being *in* authority, but without *exousia*, that physician will never heal *with* authority.

EXOUSIA AND VIRTUE

The Greek terms *dynamis* and *exousia* correspond to the Latin terms *potestas* and *auctoritas*, respectively. The Romans used *potestas* to describe the rule of Nero and Caligula, but *auctoritas* to describe the rule of Caesar and Augustus ([15], p. 52). The Western concept of rights as powers, a concept which strongly influences contemporary discussions of medical ethics, developed around the concept of *potestas* ([9], p. 61). Originally, *auctoritas* or authority had a meaning similar to that of *exousia*. But as discussed above, authority assumed a definition based on the concept of power in the writings of Hobbes and Locke. Since then, it seems that the distinction between power and authority (*potestas* and *auctoritas*; *dynamis* and *exousia*) has nearly vanished from Western writing. Consequently, contemporary writings about authority in the doctor-patient relationship have been largely oriented either to assert the traditional power of the doctor over the patient or to defend a revolt in which the patient's power is asserted over against the doctor's power.

Exousia represents, to some extent, a *tertium quid* in this examination of the relationship between doctors and patients. In the New Testament understanding, authority does not originate from either the patient or from the doctor. "Like everything human, the measure of excellence in authority is its ordination to God and its success in ordaining its subjects to God" [28].

Exousia is not itself a virtue. It is not an Aristotelian mean in the sense of being the just equilibrium point between the opposed poles of excessive control for either the doctor or the patient. *Exousia* is both an orientation to virtue and the fruit of virtue. *Exousia* results from the recognition by both the doctor and the patient that their relationship is not oriented to one or another of two individual human beings, but to a "third thing" (i.e., to God).

Exousia is an orientation to a *telos*. It is the recognition of the *telos* and the subordination of all related activities to the *telos*. As such, *exousia* is both the orientation to virtue and the possibility of virtue. Without a *telos*, there is no virtue. To speak of virtue is to presume the authority of an excellence towards which virtuous activities are oriented. Virtue demands the recognition of authority. And, once one acknowledges an authentic *telos*, one's actions are expected to be virtuous.

Exousia may be likened to grace. One does not earn or own *exousia*. Yet, it can be expressed and it can be recognized. But it cannot emerge unless its divine source has been recognized. And unless it bears fruit in virtuous life and points beyond itself, any claim to *exousia* is disingenuous.

To practice medicine with *exousia* is to ordain one's practice to the good of the patient and to ordain one's practice for the good of the patient to the glory of God. In this way, the *dynamis* to heal, which is already given in nature and in human reason, not only becomes actual but has a context and an ultimate orientation, emerging from God and leading back to God. *Exousia* therefore demands the virtues of practice: wisdom, equanimity, selflessness, trustworthiness, concern, and fidelity. The role of the doctor is defined by an oath to practice in keeping with the virtues demanded by God's free gift of the *exousia* to care for the needs of the sick. A physician practices *with* authority to the extent that this oath is upheld [39].

Virtue is also expected of the patient, but healing is never withheld because a patient does not live up to the perfect fulfillment of these virtues. The patient must also realize that the grace of healing is mediated through flawed and fragile human beings who may not live up to the virtues demanded by *exousia*. The virtues of the patient concern the stewardship of the body, which is given as a gift by God. Patients can be asked to care for their bodies, to avoid what is harmful to their bodies, to be compliant with prescriptions, and to be honest historians. But even the good of the body must be subordinated to the *telos*, which transcends the body itself.

THE WISDOM OF BEN SIRA

The deuterocanonical text of Ben Sira (the Book of Sirach or Ecclesiasticus) is included in the wisdom literature of the Roman Catholic Scriptures and is referred to 82 times in the Jewish Talmud ([37], pp. 17–20). The physician's poem from this text [38:1–15] helps to provide insights into the view of authority in the healing relationship within the Judeo-Christian tradition.

In ancient Hebrew thought, healing was traditionally reserved for God alone. To make a claim to be able to heal, then, was to ascribe to oneself qualities traditionally reserved for God alone, thus making oneself God's equal. This was an abomination. It was among the worst of all sins. It was the practice of magic and darkness associated with the enemies of God: the herbs and spells and incantations of idolaters [40].

But the rational medicine of the Greeks was not only wiser and more efficacious than the medicine of Babylon and Egypt, it made a claim to a rational basis for practice not associated with idolatry. Jews could contemplate availing themselves of the services of these Hippocratic physicians, then, if there were some theological way to reconcile this new rational medicine with the traditional understanding that healing came from the Almighty, not from human beings. Such a theological understanding is expressed in the physician's poem from the Wisdom of Ben Sira (ca. 175 B.C.). This understanding gave Hellenized Jews, for the first time in the history of Israel, an opportunity to practice medicine and ask for the assistance of physicians when sick [40].

While the poem does not use the words for power or authority directly in describing the relationship between doctor and patient, either in the original Hebrew or in the Greek translation written by Ben Sira's grandson, the themes raised by the poem deal quite explicitly with the topic. "From God the doctor has his wisdom," the poem insists. God endows the earth with all the healing herbs the doctor uses. The pure *dynamis* for healing comes originally from God, but it is through the doctor that "God's creative work continues without cease in its efficacy on the surface of the earth." Yet *dynamis* is not enough. The *exousia* to heal must also come from God. The first verse of the poem admonishes the patient to honor the physician not only because his services are "essential" (i.e., that he has *dynamis*), but also because it was God "who established his profession" (i.e., gave him *exousia*).

The orientation of medicine to God is made clear. Both the doctor and the patient are explicitly urged to pray. The physician does not falsely arrogate to himself powers over the patient which properly belong to God. And the patient does not insist on power and rights over and against the physician. Their relationship is a covenant of trust between doctor and patient authorized by the orientation of that covenant to the overarching covenant between God and all of God's people [40]. This view harmonizes with that of contemporary Christian theologians who characterize the doctor-patient relationship as a covenant [26, 35]. Power is not thought of as force of will, but as the actual possibility of healing. Authority is not force of will or the possession of specialized knowledge, but the mutual recognition by both the healer and the healed of the ultimate source of the power to heal and the ultimate source of the warrant to heal. To claim to heal by one's own force of knowledge or will is arrogant. To offer healing as a contract implies ownership of what belongs properly to God, and is thus intolerable. It is only by practicing under both covenants, with the *exousia* which God gives and which demands so much of the doctor, that the *dynamis* of expertise can become an actual act of healing for the patient.

EXOUSIA, FREEDOM, AND SERVICE

The Scriptural perspective which governs the conception of authority covered by the term *exousia* also offers an understanding of the relationship between freedom and authority which differs from contemporary usage. This perspective emphasizes the relationship of loving service to the concept of freedom as well as to the concept of authority. This perspective is highly relevant to discussions of the doctor-patient relationship.

Gunneweg and Schmithals write that the true authority of *exousia* "arises out of freedom and is based upon the possibility of rendering help as a servant" ([15], p. 21). During his final meal with his disciples, a dispute arises among them as to who is the greatest. Jesus admonishes them not to "lord it over" other people, but to fulfil what it means when it is said that those who have *exousia* over people are called their benefactors. Those in positions of true *exousia* must be servants, in imitation of Jesus, who stood among the disciples as the one who serves [Lk. 22:24–30]. Similarly, in the Gospel of John, Jesus urges the disciples to follow his example of service and wash each other's feet [Jn. 13:1–17]. Henri Nouwen, implicitly writing with an understanding of authority as *exousia*, asserts that compassion is the substance of legitimate authority ([29], pp. 40–43).

Exousia implies that authority is an assertion of the other in freedom. It is not mere *dynamis*, which is really indifferent to the will of the other, nor is it a coercive use of power, which is the assertion of personal will against the will of the other [28]. *Exousia* is authority which addresses human freedom and human reason. *Exousia* is authority which assumes a mutual orientation towards a *tertium quid*. *Exousia* is always at the service of others and their freedom. "An earthly authority which does not point beyond itself becomes demonic and will show itself as arbitrary, naked power" [28]. The life of Joseph Mengele provides a chilling example of what can happen when a medical professional distorts the authority of the profession far beyond the legitimacy and genuineness of *exousia*.

The concept of *exousia* captures a sense of human freedom which seems to have been overlooked by the Hobbesian perspective on authority, whether presented in the form of political philosophy or sociology. Hannah Arendt has written that "authority implies an obedience in which men retain their freedom" [3]. Such a statement must seem paradoxical in a culture which considers obedience and freedom as opposites. What kind of freedom is there which does not preclude obedience?

The freedom of *exousia* is the freedom which comes with liberation from self-preoccupation. It is the freedom which only loving service can bring. It is also the free acceptance of human nature with all its inherent limits, including death. It is liberation from the punishment of Sisyphus, condemned to the eternal trial of attempting to make those limits disappear [28]. It is therefore liberation from both the entrepreneurial approach to medicine often assumed by physicians and the consumerist approach to medicine often assumed by patients. Because the doctor does not own the authority to heal, the doctor cannot put healing up for sale on the market. Because the patient cannot purchase immortality, the patient need not expend all his or her human resources on a grandiose death-denying delusion.

The virtuous doctor, then, will practice with *exousia*, recognizing that healing is authorized by God, who also gives the possibility of healing in the resources of the earth and in the resourcefulness of human reason and imagination. In the covenant which exists between God and the healer, the physician must assume the virtues demanded by *exousia*, placing healing power at the service of others and at the service of their freedom. This means recognizing the dignity and freedom of the patient, and demands, in turn, a covenant between doctor and patient. *Exousia* implies the concept of authority to which Hauerwas referred when he wrote that it is not derived from knowledge or expertise, but from mastery of the practical moral skills involved in the physician's commitment to care for and never abandon the ill and the dying [16]. Likewise, to coerce, manipulate, or ignore the patient is incompatible with the spirit of practicing with *exousia*. Informed consent, then, assumes importance not as the patient's autonomous authorization of the physician's actions, but as the mutual recognition of the gifts of freedom and healing which only God bestows. Authority does not reside with the patient as something to be given to the doctor. Nor is authority something that resides with the doctor as something to be exercised over the patient. Rather, it is the result of the mutual recognition of the dignity of both doctor and patient, each reverencing the life of God in the other.

EXOUSIA, MYSTERY, AND HEALING

God is a holy mystery, and the awesome presence of God in the doctor-patient relationship ought never be ignored. But God's mystery ought never be invoked as a stopgap for our knowledge; a mere concept to define the limits of human science. When medical authority is considered only as control (practicing *in* authority), or when medical authority is considered only as the power of knowledge and expertise (practicing as *an* authority), the fundamental mystery of God's place in the healing relationship is obscured. But when medicine is practiced *with* authority (*exousia*), the holy mystery of God's healing presence opens out before both the doctor and the patient.

Robert Burt has complained that the increasing use of the courts to settle medical cases of ethical concern in advance of any anticipated actions by physicians and patients accepts the false presumptions that medical decision making is certain when it is not, and that direct conversation between the doctor and patient is to be avoided when it ought not. Patients, doctors, and hospitals turn instead "to the last bastion of unquestionable authority in our society: the Judge, the embodiment of the Law" [7]. In going to the courts, they fail to recognize both the ontological and moral ambiguity of those cases which fall at the "edges of life." They seek certitude and security where there is only uncertainty and insecurity. They seek control in situations which are fundamentally out of their control.

But the patient and doctor who recognize *exousia* know that the physician's authority is not called into question when there is no control and there is no knowledge. Those who base their authority on control and knowledge will experience these cases as threats. But *exousia* commits both doctor and patient to a recognition of the fundamental mystery of God's presence in the covenant between them. *Exousia* commits both to a recognition of the mysteries of death and limitation. The foundation of *exousia* is the transcendent, which is revealed in the immanence of sickness and death. In the midst of the powerlessness and confusion wrought by illness and death, faith and reverence replace desperation and delusion. Control slips away from one *in* authority. Expertise slips away from *an* authority. The power and the authority belong to God alone.

EXOUSIA, MEDICINE, AND THE SECULAR CITY

MacIntyre [24] despairs of the possibility of any kind of moral consensus regarding either what constitutes the good or how various goods ought to be related to one another. He argues that the concept of a profession is inherently linked to the concept of authority. He concludes that the vitality of all professions has been irrevocably destroyed because the concept of authority has lost all meaning in the wake of the loss of moral consensus.

This despair has been challenged by Pellegrino [31]. He notes that the recognition of the legitimacy of the claims of patients to act as moral agents in the doctor-patient relationship is a positive good which ought to be sustained and strengthened. But he also cautions that both patient and physician must be seen as moral agents. Physicians cannot become mere instruments of the patient's autonomous choices.

The view offered here, through the scriptural concept of *exousia*, would seem to obviate these difficulties by transforming the discussion from a debate about power for doctors and patients into a search for that *tertium quid* to which both can point as the source of authority. Paternalism is the mistaken view that authority has to do with knowledge and control which properly belong to the medical profession. "Autonomism" is the mistaken view that authority has to do with control which properly belongs to the patient, but which the patient grants to the physician only because the patient lacks the knowledge. The way of *exousia* is the "third way."

MacIntyre is certainly correct in his assessment that there is no moral consensus, let alone any religious consensus, in the Western world. One might therefore argue that the scriptural notion of *exousia* would be helpful only if one accepted MacIntyre's vision of small communities in which patients and staff shared the Judeo-Christian faith and its conception of the authority to heal. But a great many people would want no part of such communities, either because they have no faith or because their faith does not include the concept of *exousia*.

Does this imply that the concept of *exousia* is irrelevant in a pluralistic society? Does medicine have no unifying goal other than to maximize personal liberty to the extent that others are not harmed? Is total fragmentation of the profession inevitable?

I would suggest that the profession itself will ultimately resist fragmentation into distinct medico-moral practices. Granted, in the wake of intense specialization and sub-specialization, it might no longer be possible in a certain sense to talk of a single medical profession. Nonetheless, there seems to be enough unity inherent in the professional activities of contemporary Western physicians to resist fragmentation into little philosophically or religiously distinct professions. The interdependence of medical knowledge, the uniformity of the initial education, the fundamental belief in rational medicine based on scientific evidence, and the oath that physicians take to put this knowledge at the service of patients are critical unifying elements for the profession. Despite the many centrifugal forces which threaten contemporary medical practice, the fact of illness and the act of profession might still form, as Pellegrino points out [32], a secular basis for a *tertium quid* of the sort that could serve as a source of *exousia* for the relationship between doctor and patient. These would constitute integral constituents of the healing relationship, transcending the power concerns of both the doctor and the patient. Beneficence and autonomy might cease to be considered antithetical. Pellegrino and Thomasma locate a *telos* intrinsic to the practice of medicine: the good of the patient. This forms the basis, on secular grounds alone, of a new model for the doctor-patient relationship: beneficence in trust [34]. It requires a trusting relationship between doctor and patient not unlike the religious concept of a covenant. It demands virtue of both doctors and patients. Such a model of the healing relationship does not have the power of a truly Judeo-Christian model like the one developed in this paper, but it certainly has secular credibility. And even a secular notion of *exousia* would certainly provide a helpful alternative to the twin vices of physician paternalism and patient "autonomism."

A CLINICAL EXAMPLE

To illustrate, in a preliminary and sketchy fashion, how the adoption of a view of medical authority as *exousia* might affect the practice even of secular medicine, I will offer the following example.

Suppose a patient, dying of metastatic lung cancer, is placed on a morphine drip by his doctor to treat his severe pain. Suppose that this patient, quite medically sophisticated, begins to manipulate the drip rate on his own intravenous pump.

A doctor who conceived of authority as power might perceive this patient's behavior as a threat to the authority of the doctors and nurses. If somewhat enlightened, she might interpret this as a manipulative behavior in which the patient was acting out because of fear of death. Her reaction to this behavior would probably be to reassert control by "setting limits" so that the patient would understand the boundaries of proper patient behavior. She might, mercifully, increase the drip rate, but tell the patient that he could only ask for a change in drip rate once per nursing shift, and that he would be carefully watched so that he did not increase the rate on his own. Since he lacked the knowledge (power) to safely adjust the rate, she would insist that control of the morphine dose must be the prerogative of the doctor. If he objected, mutual anger and a stalemate might ensue.

On the other hand, if the doctor conceived of authority as *exousia*, she might behave differently. She might ask herself, and the patient, what the *telos* was at this stage in the illness and in their doctor-patient relationship. She might ask how they could work towards the overall *telos* of medicine (the good of the patient) in this situation, and how the patient saw his ability to control the rate of the morphine drip fitting into that *telos*. They might agree that the power of the drug (a power which belongs to neither of them) could be better expressed in the service of that *telos* if the patient actually could manipulate the dose of morphine within certain bounds of safety. Thus, the goal both shared could be achieved and the power of medicine more fully expressed in the setting of their mutually trusting relationship.

One can only speculate that it was such an exchange which must have given rise to the wonderful new development of Patient Controlled Analgesia (PCA), in which the patient is given a baseline infusion rate of narcotic but can give additional drug boluses (limited by safety concerns) as necessary to relieve pain not relieved by the baseline infusion. The sum of the boluses required are then used to calculate adjustments in the baseline rate [42].

Some might object that the use of PCA for such a patient merely illustrates an effective redistribution of power and control from doctor to patient, but this interpretation would miss the point. Certainly it is not patient power or doctor power which is at stake here. The things that matter are mutual trust, shared goals, and the expression of a power which belongs not to the doctor or to the patient, but to Humankind or to Nature or to God (or to all three). Certainly it is not control, but loss of control which is the dominant theme in such a situation. To engage in a struggle for control over the rate of morphine infusion as an expression of continued personal control in the face of the overwhelming reality of the patient's imminent, ineluctable death would be absurd for both doctor and patient. The wise physician recognizes the dignity and freedom of her patient even in these final moments, and both doctor and patient ordain their freedom to the common goal of alleviating suffering even in the face of their mutual powerlessness to prevent death. Thus would a wise physician practice *with* authority, in the spirit of *exousia*.

CONCLUSION

Contemporary discussions of authority in the doctor-patient relationship have largely been based on either sociological or political models. Both of these models assume that authority means control or expertise or both. Under such models, the notion of authority in the doctor-patient relationship has been viewed as a struggle for power between doctors and patients. This perhaps has generated some of the difficulties now confronting medicine, where many physicians continue to exert paternalistic control over patients while many patients now practice medical consumerism. The New Testament view of authority, *exousia*, provides an alternative view which suggests that the authority to heal is neither a possession of the physician nor of the patient, but a free gift from God. *Exousia* is the warrant to heal. *Exousia* requires virtue from both doctor and patient in a covenant relationship, subsumed under the greater covenant between God and all God's people. The notion of *exousia* also points to a way for secular medicine to move beyond the contradictions of medical authority conceived of as knowledge and control. Perhaps a recovery of the notion of *exousia* can help medicine escape from a spirit of antagonism which is increasingly making the experiences of both going to the doctor and of practicing medicine unsatisfactory for both doctors and patients.

ACKNOWLEDGEMENTS

I am grateful to Dr. Edmund D. Pellegrino for his thoughtful review of a draft of this manuscript and to the Charles E. Culpeper Foundation for their generous support of my work.

Center for Clinical Bioethics
Georgetown University Medical Center
Washington, DC
U.S.A.

NOTE

[1] It is of interest that while Locke was trained as a physician, he never saw any patients professionally.

BIBLIOGRAPHY

1. Amiot, F., and Galopin, P. M.: 1973, 'Authority', in X. Leon-Dufour (ed.), *Dictionary of Biblical Theology*, 2nd ed., Seabury Press, New York, pp. 36–39.
2. Aquinas, St. Thomas: 1966, *Summa Theologiae*, vol. 28, Blackfriars edition, T. Gilly, O.P. (ed.), McGraw-Hill, New York.
3. Arendt, H.: 1968, 'What Is Authority?', in *Between Past and Future*, Penguin Press, Middlesex, England, pp. 91–141.
4. Baier, A. C.: 1987, 'The Need For More Than Justice', in M. Hanen and K. Nielsen (eds.), 'Science, Ethics, and Feminism', *Canadian Journal of Philosophy* 13 (Suppl.), 41–56.
5. Beauchamp, T. L.: 1982, *Philosophical Ethics*, New York, McGraw-Hill.
6. Beauchamp, T. L., and Childress, J. F.: 1989, *Principles of Biomedical Ethics*, Oxford University Press, New York.
7. Burt, R. A.: 1988, 'Uncertainty and Medical Authority in the World of Jay Katz', *Law, Medicine, and Health Care* 16, 190–196.
8. Engelhardt, H. T.: 1986, *The Foundations of Bioethics*, Oxford University Press, New York.
9. Engelhardt, H. T.: 1991, *Secular Humanism: The Search for a Common Morality*, Trinity Press International, Philadelphia.
10. Faden, R. R., and Beauchamp, T. L.: 1986, *A History and Theory of Informed Consent*, Oxford University Press, New York.
11. Finnis, J.: 1980, *Natural Law, Natural Rights*, Clarendon Press, Oxford, England.
12. Flathman, R.: 1982, 'Power, Authority, and Rights in Medicine', in G. J. Agich (ed.), *Responsibility in Health Care*, D. Reidel, Dordrecht, the Netherlands, pp. 105–125.

13. Foerster, W. F.: 1964, 'Exousia', in G. Kittel, and G. Friedrich (eds.), *Theological Dictionary of the New Testament*, vol. II, Eerdmans, Amsterdam, the Netherlands, pp. 562–575.
14. Friedman, R. B.: 1991, 'On the Concept of Authority in Political Philosophy', in J. Raz (ed.), *Authority*, New York University Press, New York, pp. 56–91.
15. Gunneweg, A. H. J., and Schmithals, W.: 1982, *Authority*, J. E. Steely (trans.), Abdingdon Press, Nashville, Tennessee.
16. Hauerwas, S.: 1982, 'Authority and the Profession of Medicine', in G. J. Agich (ed.), *Responsibility in Health Care*, D. Reidel, Dordrecht, the Netherlands, pp. 83–104.
17. Hobbes, T.: 1651 (1946), *Leviathan, or the Matter, Forme, and Power of a Commonwealth, Ecclesiastical and Civil*, M. Oakshott (ed.)., Basil Blackwell, Oxford, England.
18. Kierkegaard, S.: 1955, *On Authority and Revelation: The Book on Adler*, W. Lowrie (trans.), Princeton University Press, Princeton, New Jersey, p. XVI.
19. Ladenson, R.: 1991, 'In Defense of a Hobbesian Conception of Law', in J. Raz (ed.), *Authority*, New York University Press, New York, pp. 32–55.
20. Lavin, B., Haug, M., Belgrave, L. K., and Breslau, N.: 1987, 'Change in Student Physicians' Views on Authority Relationships with Patients', *Journal of Health and Social Behavior* 28, 258–272.
21. Locke, J.: 1690 (1988), 'An Essay Concerning The True Original Extent and End of Civil Government', in *Two Treatises of Government*, P. Laslett (ed.), Cambridge University Press, New York, pp. 256–428.
22. Mackie, J. L.: 1977, *Ethics: Inventing Right and Wrong*, Penguin Books, Harmondsworth, Middlesex, England.
23. MacIntyre, A.: 1967, *Secularization and Moral Change*, Oxford University Press, London, England.
24. MacIntyre, A.: 1977, 'Patients as Agents', in S. F. Spicker, and H. T. Engelhardt (eds.), *Philosophical Medical Ethics: Its Nature and Significance*, D. Reidel, Dordrecht, the Netherlands, pp. 197–212.
25. MacIntyre, A.: 1984, *After Virtue*, University of Notre Dame Press, Notre Dame, Indiana.
26. May, W. F.: 1983, *The Physician's Covenant*, Westminster Press, Philadelphia.
27. Meyers, A. C. (ed.): 1987, *The Eerdman's Bible Dictionary*, Eerdmans, Grand Rapids, Michigan, pp. 108; 844–845.
28. Molinski, W.: 1975, 'Authority', in K. Rahner (ed.), *The Concise Sacramentum Mundi*, Seabury Press, New York, pp. 60–65.
29. Nouwen, H. J.: 1972, *The Wounded Healer*, Doubleday, Garden City, New York.
30. Parsons, T.: 1951, *The Social System*, The Free Press, Glencoe, Illinois.
31. Pellegrino, E. D.: 1977, 'Moral Agency and Professional Ethics: Some Notes on the Transformation of the Physician-Patient Encounter', in S. F. Spicker, and H. T. Engelhardt (eds.), *Philosophical Medical Ethics: Its Nature and Significance*, D. Reidel, Dordrecht, the Netherlands, pp. 213–220.
32. Pellegrino, E. D.: 1979, 'Towards a Reconstruction of Medical Morality: The Primacy of Act of Profession and the Fact of Illness', *Journal of Medicine and Philosophy* 4(1), 32–56.
33. Pellegrino, E. D., and Thomasma, D. C.: 1981, *A Philosophical Basis of Medical Practice*, Oxford University Press, New York.
34. Pellegrino, E. D., and Thomasma, D. C.: 1989, *For The Patient's Good: Towards the Restoration of Beneficence in Medical Ethics*, Oxford University Press, New York.
35. Ramsey, P.: 1970, *The Patient As Person*, Yale University Press, New Haven, pp. xi–xviii.
36. Raz, J.: 1991, 'Introduction', in J. Raz (ed.), *Authority*, New York University Press, New York, pp. 1–19.
37. Skehan, P. A., and DiLella, A. A.: 1987, *The Wisdom of Ben Sira*, Anchor Bible Series, vol. 39, Doubleday, New York.
38. Suchman, A. L., and Matthews, D. A.: 1988, 'What Makes the Doctor-Patient Relationship Therapeutic? Exploring the Connexional Dimension of Medical Care', *Annals of Internal Medicine* 108, 125–130.

39. Sulmasy, D. P.: 1989, 'By Whose Authority: Emerging Issues in Medical Ethics', *Theological Studies* 50, 95–119.
40. Sulmasy, D. P.: 1989, 'The Covenant Within the Covenant: Doctors and Patients in Sirach 38:1–15', *Linacre Quarterly* 55(4), 14–24.
41. Sulmasy, D. P., Geller, G., Levine, D. M., and Faden, R.: 1990, 'Medical House Officers' Knowledge, Confidence, and Attitudes Regarding Medical Ethics', *Archives of Internal Medicine* 150, 2509–2513.
42. Tansen, A., Hartvig, P., Fagerlund, C., and Dahlstrom, B.: 1982, 'Patient-controlled Analgesic Therapy: Part II. Individual Analgesic Demand and Plasma Concentrations of Pethidine and Post-operative Pain', *Clinical Pharmacokinetics* 7, 164–175.
43. Veatch, R. M.: 1982, 'Medical Authority and Professional Medical Authority: The Nature of Authority in Medicine for Decisions by Lay Persons and Professionals', in G. J. Agich (ed.), *Responsibility in Health Care*, D. Reidel, Dordrecht, the Netherlands, pp. 127–137.
44. Veatch, R. M.: 1981, *A Theory of Medical Ethics*, Basic Books, New York.
45. Weber, M.: 1947 (1968), *Theory of Social and Economic Organizations*, A. M. Anderson and T. Parsons (trans.), Free Press, New York.
46. Wolff, R. P.: 1991, 'The Conflict Between Authority and Autonomy', in J. Raz (ed.), *Authority*, New York University Press, New York, pp. 20–31.

JOHN W. GLASER

CONFLICTING LOYALTIES: BENEFICENCE – LOVE WITHIN LIMITS

The purpose of theology is to serve the fullness of community life. One service of this discipline is to provide us with conceptual models that help us reflect and clarify central issues with direction and consistency. These qualities can then flow into action of individuals and the larger community. In terms of clinical ethics this means that the conceptual systems developed by theological analysis can be expected to be useful in the practical doing of ethics.

I believe that a Christian tradition that considers beneficence as the center of the moral life has much to offer the current discussion of health care ethics. In much recent bioethical discussion "beneficence" is only one of several basic principles. But in a long-standing Christian tradition beneficence stands as the foundational principle, grounding all else. I believe that this classic Christian perspective on the moral life can open some fresh vistas for the ethics of health care.

In this essay I will first develop a basic model of the moral life as beneficence-grounded. I will then expand this fundamental idea of beneficence beyond the realm of interaction between individuals, to include realms of "institutional and societal beneficence." And finally I will explore some implications of this expanded beneficence model.

PART ONE: A SKETCH OF ETHICS AS BENEFICENCE –
LOVING IN THE SITUATION OF FINITUDE

An adapted parable captures the heart of the problem we will explore.

The Conflicted Samaritan:
A man was once on his way down from Jerusalem to Jericho and fell into the hands of brigands; they took all he had, beat him and then made off, leaving him half dead. Now a priest happened to be traveling down the same road, but when he saw the man, he passed by on the other side. In the same way a Levite who came to the place saw him, and passed by on the other side. But a Samaritan traveler who came upon him was moved with compassion when he saw him. He went up and bandaged his wounds, pouring oil and wine on them. He then lifted him on his own mount. As the Samaritan traveled further he came upon another man who had been beaten and needed care. He likewise ministered to him and set him on his mount. As the three turned the next bend in the road the Samaritan's heart sank for there were two more figures lying on the side of the road in the foreground and further, before the road turned in the distance he made out one further traveler, struck to the ground and needing help. His heart was filled with pity and compassion – but with growing distress – for his resources would be exhausted long before he reached the last person in his view. And he could only guess at what lay around the next bend.

G.P. McKenny and J.R. Sande (eds.), Theological Analyses of the Clinical Encounter, 109–132.
© *1994 Kluwer Academic Publishers. Printed in the Netherlands.*

This parable presents two kinds of ethical conflicts:
1. the *existential ethical crisis* of the priest, Levite, and Samaritan who confront their duty and make a decision. We will not concern ourselves with such ethical decisions; and
2. the *normative ethical crisis* where in Dewey's terms: "The struggle is not between a good which is clear to him and something else which attracts him but which he knows to be wrong. It is between values each of which is an undoubted good in its place but which now get in each other's way" ([10], p. 7).

With this parable as a starting point I want to argue that all normative decisions represent such conflicting loyalties and that we have not taken sufficient note of this underlying values-in-conflict fabric of our ethical life.

Love of neighbor stands as the basic commandment of Christian morality. In *Quadragesimo Anno* Pius XI says: "all the commandments . . . may be reduced to the single precept of true charity" ([30] #137).

In Christian Scripture the most dramatic and stark presentation of the central role of love of neighbor comes in Mt. 25 when religion and morality are cast in these surprisingly simple terms of response to neighbor's need.

Then the King will say to those on his right hand, "Come, you whom my Father has blessed take for your heritage the kingdom prepared for you since the foundation of the world. For I was hungry and you gave me food; I was thirsty and you gave me drink; I was a stranger and you made me welcome; naked and you clothed me, sick and you visited me, in prison and you came to see me." Then the virtuous will say to him in reply, "Lord, when did we see you hungry and feed you; or thirsty and give you drink? When did we see you a stranger and make you welcome; naked and clothe you; sick or in prison and go to see you?" And the King will answer, "I tell you solemnly, in so far as you did this to one of the least of these brothers of mine, you did it to me" [Mt. 25:34–40].

Karl Rahner has repeatedly elaborated the theology behind this passage when he delineates the radical unity of love of God and love of neighbor [31, 32]. He poses the question: is love of neighbor identical with love of God to the radical extent that no act of love of God can occur which is not an act of neighbor-love? His answer: "The primary act of love of God is the act of categorial-explicit love of neighbor. In this act of neighbor-love God is directly met in supernatural transcendentality – always unthematic but actually. The explicit act of love of God is always borne by this trusting, loving opening to all of reality which occurs in the act of love of neighbor. It is true with a *radical* – not merely psychological or 'moral' – necessity, that one who does not love one's brother, whom one 'sees,' cannot love God, whom one does not see, and that a person can only love God, whom one cannot see *in so far* as a person loves their visible neighbor" ([31], p. 295).

If the commandment to love our neighbor is foundational, what of the other commandments – not to steal, kill, covet, and the like? How do they relate to this fundamental commandment? As species to genus. They translate the foundational commandment of love into specific areas of life. Paul's letter to the Romans spells this out.

If you love your fellow men you have carried out your obligations. All the commandments: *You shall not commit adultery, you shall not kill, you shall not steal, you shall not covet*, and so on, are summed up in this single command: *You must love your neighbor as yourself*. Love is the one thing that cannot hurt your neighbor; that is why it is the answer to every one of the commandments [Romans 13:8–10].

Such passages have led systematic theologians such as Bruno Schüller to identify love as the foundation and love's structure as determinative of morality's method: "The double commandment of love of God and love of neighbor encompass the totality of moral claims that can be made on a person. Therefore, the manner of reasoning that serves to clarify this double commandment must be seen, in all cases, as fundamental" ([35], p. 527, pp. 34–42). Another Catholic theologian, R. Carpentier, identifies the essence of theological ethics to consist in articulating "the thousand fold specific and concrete formulations of charity, the mother and root of all virtue" ([7], pp. 53–54). In carrying out this task of the thousand fold specification of love further distinctions are necessary.

Love as Benevolence/Beneficence

Catholic tradition has understood charity as comprised of two distinct but complementary and necessary elements: love as *benevolence* and love as *beneficence*. It is not enough to wish our neighbor well (bene velle), we must also act for their good (bene facere)."

Charity is first and foremost benevolence – an attitude of the mind and heart that wishes the best for one's neighbor. Such an attitude is the heart of charity, the roots from which deeds of love spring. Love *as benevolence* is both comprehensive and universal. It wishes all good things to the neighbor and it recognizes the neighbor as everyone – including one's enemies. Benevolence knows no limits, it reaches, like God's love, from end to end with might and embraces all hearts with tenderness.

With love *as beneficence* things are leaner, more severe. Beneficence is wholeheartedly directed to doing good but *beneficence knows severest limits*: limits of knowledge, imagination, time, space, ability, resources. Essential to beneficence is its characteristic of love-within-limits. As Gustafson puts it: "the good is sought under the conditions of finitude" ([18], p. 141). It is our nature as finite creatures, not a narrowness of heart, that fundamentally accounts for the limits of beneficence. It is precisely this characteristic of whole-hearted love, but *love within limits*, that sets a major agenda for an ethics of beneficence.

It is obvious that beneficence as understood in this way is essentially different from beneficence as it is commonly understood in the recent literature of bioethics. The latter understanding is much narrower than the former and is virtually synonymous with a paternalistic attitude toward patients. Pellegrino and Thomasma comment: "This is the conception of beneficence still dominant in the minds of many physicians and patients; it still shapes the ethos and ethics

of medicine. It is the conception, too, that is the focus of criticisms by proponents of autonomy who equate beneficence almost entirely with medical
paternalism" ([29], p. 13).

The understanding of our moral life proposed in this essay claims that we
are always called to fulfill the basic commandment of love, of beneficence.
And the nature of beneficence involves choosing between limited options.
To illustrate this Schüller offers the example of a physician who stands in
the situation where she can only provide a benefit to her patient by causing
pain. Schüller then makes this point – a point foundational for our considerations here:

In such a situation one stands before two values that compete with one another. To realize one
value a person must leave the other unrealized. One must then decide which of the values deserves
priority. If we look carefully we see here the characteristic and fundamental human condition:
as a limited being, a person has only limited possibilities available for serving the neighbor's
good. A person's actions cannot effectively benefit everyone, nor respond to any and every
legitimate need and deficit. One must make a choice and decide which of those currently
available possibilities deserves to be selected. One must, therefore, be able to choose between
important and more important, between urgent and less urgent, between better, good and less
good. This means that one must be able to grasp the preference imperative that prevails here
amidst competing values. In our daily lives we are not normally conscious of this constant
choosing between competing values. In any case, we have become so accustomed to this
unrelenting value preference that it hardly ever catches our attention ([42], p. 70).

The moral imperative – the beneficence imperative – emerges from the
weighing of human goods to be realized and evils to be avoided. Schüller
makes two points central to our discussion:
1. this situation of facing and having to choose between conflicting values/
   disvalues constitutes the essential fabric of our moral life ("If we look
   carefully we see here the characteristic and fundamental human condition");
2. we are usually unaware of this situation ("we have become so accustomed to this unrelenting value preference that it hardly ever catches our
   attention").
These two points parallel Max Weber's comment, cited by Schüller, that "all
action . . . in its consequences boils down to partisanship on behalf of some
values and – a fact so readily overlooked today – against other values" ([35],
p. 532).

What I have described in terms of beneficence can also be viewed from
other partial perspectives and described in terms of values, duties, rights,
loyalties, virtues, principles, etc. I want to shift the perspective slightly and
summarize our discussion up to this point in terms of *ethical principles*.

In terms of ethical principles the architecture of such a beneficence ethics
would be structured as follows:
1. Beneficence – not one of many coordinate principles but the foundational
   principle of which all else is specification. It is grounded in benevolence
   – that attitude of heart that wishes well to all.
   1.1. Principle of proportionality – not one of many principles but the
        grounding *instrumental* principle which applies beneficence to specific

situations. It can be referred to in other terms – as the principle of reasonableness, the best interest principle, or benefits-burdens principle.

1.1.1., etc.

>All other principles – respect for life, respect for death, truth telling, respect for autonomy, informed consent, social justice, etc. – are principles with no pre-ordained hierarchy; the priority must be established by ethical discernment in each specific situation.

An Overlooked Constant of Ethics

The insight that the fundamental situation of normative ethics always involves a choice for some values to the detriment of others has found various expressions across millennia and traditions. Cicero, Augustine, Aquinas, Max Weber, Lessing, Hartmann, Fichte, Ross, Thielicke, Dewey – among others speak to this issue. Still, in my opinion, its cardinal importance has not been recognized and exploited for healthcare ethics. This is puzzling, and deserves some extended discussion before we move on.

Prominent authors – rather than identifying values-in-conflict as the fabric of ethical existence – explicitly identify this as an *occasional characteristic* in ethical experience. For example, we read: "Sometimes we confront two or more prima-facie duties or obligations, one of which we cannot fulfill without sacrificing the other(s)" ([8], p. 429). Another author says: "It is clear that an increasing number of theologians insist on understanding moral norms within the conflict model of human reality. Conflicted values mean that occasionally our choices (actions or omissions) are inextricably associated with evil. Thus we cannot always successfully defend professional secrets without deliberately deceiving others. . . ." ([27], p. 75). If, as I am claiming, such value conflicts are omnipresent, how might one explain that this universal situation and its importance for ethics are so little noticed?

A partial explanation lies in Schüller's citation of Wittgenstein suggesting that "the aspects of things that are most important to us are concealed under their simplicity and their everyday nature." ("One cannot notice the thing because it is always right in front of our eyes" [40], p. 650.) We can compare this to our breathing, to the structure of our mother tongue, to the rules of logic – we are inclined to notice these "infrastructures of life" only when they take the forms of aberrations or exaggerations. The *usual* presence of these realities is an unthematic rhythm of life, remaining beneath the threshold of consciousness. Like the law of gravity it is always present but not reflexively conscious.

Then too, most of the daily value conflicts are so disproportionate that their resolution is obvious and requires no explicit attention. Overlapping this consideration is the fact that macro life decisions – choice of profession, marriage, parenthood, and so forth – imply circles of subordinate micro-decisions that flow spontaneously from the priorities established in the macro-decision. Again, the resolution of countless value conflicts takes place below the threshold of our attention.

Further, our language usage and mental paradigms habitually direct our attention away from rather than toward this value conflict dimension of life. I think of an experience I had recently which involved actions that could in certain circumstances qualify as: battery, sexual assault, invasion of privacy, infliction of bodily pain, causing anxiety, and inflicting financial loss. How did I refer to this experience in my conversations? As my annual physical exam. That is, I name this actual conflict of values in terms of the preferred value(s) that I judge to outweigh the disvalues involved. The same could be said of terms such as medication, hospitalization, inoculation, chemotherapy, surgery, X-ray, blood work, and so on, *ad infinitum*.

It is important to note how much language goes beyond being a mirror of life and functions as hammer and anvil of our conscious experience. Language shapes consciousness certainly as much as it reflects it. Werner Stark suggests that we recognize the role of language as a mental grid. "We see the broad and deep acres of history through a mental grid . . . through a system of values which is established in our minds *before* we look out onto it – and it is this grid which decides . . . what will fall into our field of perception" ([43], p. 16, pp. 7–8).

Sometimes it is our very ethical/philosophical efforts that do us the disservice of concealing this value-conflicting dimension of life. For example, the "principle of double effect" is part of the problem. What we have elaborated above says: all normative issues involve actions with not only double but multiple effects. To single out – relatively rare – cases where "the principle of double effect" is applicable clearly implies that those other cases – the majority – do not have to be concerned about such conflicting value issues.

Another example is the oft repeated phrase from Hippocratic ethics: *primum non nocere* – first of all, do no harm. This phrase hides the *basic operational mode of all medical intervention*: doing harm because greater good is expected from it. As I pointed out above, all medical care visits harms on patients – use of time, needle sticks, chemical insults, diminishment of money, and so forth. Health care without harm is nonsense. The revealing rather than concealing formulation of the principle might read: *nocete proportionaliter* – do harm to the extent that it is justified by the promise of greater good.

A further fogging element from the ethical enterprise itself is the overemphasis on the differences between teleology and deontology and a corresponding neglect of the deeper and more fundamental problem that each of them must face: how to resolve the conflicts that inevitably arise regardless of whether our basic focus is rules or actions, "duties" or "consequences." Veatch refers to this when he says: "Utilitarians, formalists, and mixed formalists all have the problem of resolving competing ethical claims" ([47], p. 10).

Beyond this, the very terms "Christian charity" and "beneficence" diminish rather than heighten the awareness of conflict. Charity conjures up images of a state of frictionless harmony rather than a struggle with hard choices that result in some individuals and groups necessarily facing negative, possibly very harsh, consequences. Paul's words can lull us into the belief that hard choices are beyond charity: "Love is always patient and kind; it is never jealous; love is never boastful or conceited; it is never rude of selfish; it does not take offense, and is not resentful. Love takes no pleasure in other people's sins but delights in the truth; it is always ready to excuse, to trust, to hope, and to endure whatever comes" [1 Cor. 13:4–7.].

Also, ethics often speaks in a kind of voice – which Gustafson calls the "prophetic ethical voice" [18] and which Schüller – borrowing from Stoic philosophy – refers to as Päranese ([42], pp. 14–40). This is a manner of ethical discourse that assumes that we know *what* should be done and gets on with *the doing* of what is good through encouragement, inspiration, warning. Virtually all New Testament ethics speaks in this voice. In the words of Schreiner, such ethical discourse "arrives with penetrating force of expression, lacking all mitigating qualification, with a naked, honed edge of command. Its sentences are intended not as formulations for casuistic referral, but as clarions to startle attention" ([33], p. 19). Such an apodictic and prophetic voice operates much like the term "medical examination" as noted above. It points to a stage beyond normative discernment, where conflicts have been resolved and norms have been established.

Mary Ann Glendon reminds us that as Americans we have a further cultural-linguistic handicap in perceiving such value conflict [16]. She argues extensively that as Americans we are addicted to court-oriented rights talk as a way of perceiving and talking about life. According to her, no other nation is so thoroughly addicted, nor does any other society define it in terms as fiercely adversarial, as ours. The result of this is an all-or-nothing, yes-no, win-lose mentality. Such patterns of thought, affect, and conversation tend to severely reduce the parameters of discussion, to emphasize the conclusion and its correctness, and to experience competing considerations not as counterbalancing and qualifying truths, but as error to be dismissed.

Even a cursive list of considerations such as the above make it understandable that although we are always choosing between competing goods/values/loyalties we only become aware of this by way of the rarest exception. A primary task of an ethic of beneficence is to provide this conflictual dimension of beneficence and the key elements of its resolution the prominent place that they deserve.

PART TWO: AN EXPANDED CONCEPT OF BENEFICENCE – CONCENTRIC CIRCLES
OF INDIVIDUAL, INSTITUTIONAL, AND SOCIETAL BENEFICENCE

The complexity of this is further compounded by the fact that we not only choose between conflicting goods concerning individuals, but also within and between concentric realms of values/goods/loyalties: the individual realm, the organizational realm, and the societal realm. The U.S. Catholic bishops say: "Christians believe that Jesus' commandment to love one's neighbor should extend beyond individual relationships to infuse and transform all human relations from the family to the entire human community. . . . Such action necessarily involves the institutions and structures of society, the economy and politics" ([45], p. 732). Schematically we can imagine a model of beneficence along the lines sketched below.

Realms of Beneficence/Ethics

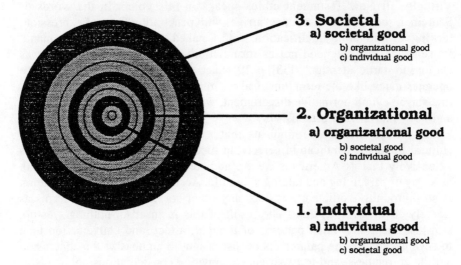

3. Societal
a) societal good
b) organizational good
c) individual good

2. Organizational
a) organizational good
b) societal good
c) individual good

1. Individual
a) individual good
b) organizational good
c) societal good

Individual Beneficence/Ethics

The simplest realm of beneficence/ethics is the realm of individual beneficence. Here the concern is primarily with *the good of individuals*. This concerns individuals and their relationships: the relationships that exist *within one individual* between various values and needs – physical, emotional, mental, and spiritual. It also attends to differences in degree and intensity within and between these goods – for example it must weigh the relative importance of intense physical good and moderate spiritual good. It attends to differences of probability and certainty, for example, between near certain emotional harm of a moderate degree and probable intellectual benefit to an extensive degree. It must attend to the whole range of comparable elements such as long/short term, partial/total, transient/abiding, direct/indirect, central/peripheral.

This realm also deals with weighing and balancing the values/goods/ loyalties that stand in tension *between two or more individuals*. For example, we must weigh my privacy and your need to have information about me, or the need of one person to be treated and the danger of infection for the professional providing treatment. Again the issues of probability, long/short term trade offs, degree and extent of harms and benefits all come into play.

The first two decades of bioethics have dealt extensively with this realm of 1a. Most of this era's burning questions fit comfortably in this realm of individual good: patient autonomy, informed consent, privacy, patient rights, truth-telling, and confidentiality.

But beyond the intra- and inter-individual issues are questions that treat sphere 1b): relationships of individuals to organizations. What responsibilities do patients, nurses, physicians have to their hospital? What trade-offs in income, safety, efficacy of treatment, and confidentiality can individuals be expected to make for the benefit of the institution?

Beyond this realm are issues in the sphere of 1c): relationships of individuals to the common good of society. What personal benefits should I forego or burdens should I bear in order to make community benefits available or harms avoidable? For example, what limits on care, what delays or diminished quality should an individual accept in order that the whole community can be assured of basic services?

So, in this realm of individual beneficence/ethics there are three sub-perspectives: a) primarily within and between individuals; secondarily b) from individuals toward organizations and c) from individuals toward the larger society.

Organizational Beneficence

Normally, the use of the word beneficence has only individuals as its referent. The present analysis understands beneficence in terms of organizations as well. Organizations are both subject and object of beneficence.

The social realities that I refer to as "organizations" – a family, a union, a business, a hospital, a religious community – have an identity, a purpose, a history and character. They have vital systems which account for their vigor and health. They have commitments, claims, relationships, and responsibilities.

The primary object of organizational beneficence is the *net organizational good* – that is, a state of organizational vigor and development that enables the organization to maximize its purpose now and into the future. Each organization has its seat of responsibility – parents in a family, officials in a religious community, governance and management in a hospital. They must seek the net good of the organization as individuals seek net good at an individual level. But obviously the resolution of beneficence choices, in terms of complexity and extent, increases exponentially at this level.

But such pursuit of the organizational good must also consider: 2c) the

individual good of those within the organization. For example, let us assume
a demonstrated need for the good of a hospital to reduce its size. There are
usually many ways to accomplish such a goal. The imperative of beneficence
is to find the complex balance of burden/benefit distribution that serves
organizational net good, but also attends to the needs of individuals. At what
point, if any, does *this* kind and *this* intensity of individual need outweigh
that kind and *that* intensity of organizational need? Does the sheer number
of individuals involved – and if so, how many – shift the balance in weighing
alternatives?

Organizational beneficence must also attend to 2b), the common good of
the society within which the organization exists. For health care institutions
not only provide health services, they are also a powerful cultural force and
agent. By their presence, their promotional efforts, their budgets and their
services health care institutions have a significant influence on what the general
population thinks, hopes and demands in terms of health care. Hospitals not
only respond to but also create demand in the general public about what to
expect of a hospital by way of service, convenience, and opulence. Hospitals
shape voter apathy, energy, and indignation – subtly but powerfully. Beyond
such indirect political influence, health care organizations also exert direct
political force on the larger system of quality, access, and financing.

Beneficence in the sphere of 2b) sustains a consciousness of the organization's impact on the larger society and insists that as the organization pursues
its own net good it do so constrained by this consideration: how can we best
achieve the net good of the organization while also promoting the common
good of society?

In daily operations most of these issues are thought of as "operational
questions," "organizational issues," "financial concerns," "management issues,"
or "marketing programs." They are that. But in the terms of this essay they
must also be identified as central issues of organizational beneficence and
therefore, vital issues of ethics.

Societal Beneficence

The final realm of an ethic of beneficence is that of society. This realm deals
with the common good of society. The Hastings Center Report defines the
common good as "that which constitutes the well-being of the community –
its safety, the integrity of its basic institutions and practices, the preservation
of its core values." It also refers to the *telos* or end toward which the members
of the community cooperatively strive – the "good life, human flourishing, and
moral development" ([21], p. 6; [3, 46]). Garrett Hardin offers a helpful illustration of the common good and how it differs and even conflicts with the good
of individuals. He asks us to think of a group of herdsmen who share a common
grazing pasture. As long as there is enough pasture to feed the cattle and
rejuvenate itself for the future, each individual herder can pursue personal
aggrandizement without jeopardizing the common good. But at some point

the danger of overgrazing emerges if each individual continues to increase the size of his herd. As long as the horizon of reflection remains individual – "what benefit comes *to me* from adding one more animal to my herd?" – the problem can neither be identified in a timely way nor resolved. Hardin says: "Therein is the tragedy. Each man is locked into a system that compels him to increase his herd without limits – in a world that is limited. Ruin is the destination toward which all men rush, each pursuing his own best interest in a society that believes in the freedom of the commons. Freedom in a commons brings ruin to all" ([19], p. 1224).

Societal beneficence attends to the commons. It knows that the common good is not achieved by some invisible hand as we each pursue our own individual good. Societal beneficence brings heart, mind, imagination and hands to the nurturing of this common good.

Attending to the commons means balancing conflicting needs/goods/ loyalties – education, housing, defense, health care, art, infrastructure, and so forth. *Being unable to meet any one or all of these societal needs fully we seek a reasonable balance among them.* The major task of societal beneficence is continually to attend to this balance – correcting historical aberrations, adjusting to new forces and circumstances – so that society can be humane and nurture growth.

A further dimension of societal ethics concerns the balance within each sector of the common good – education, health, housing, and so forth. For example, achieving the "healthcare good" of society involves finding the appropriate balance among competing healthcare needs such as prevention and cure, acute and chronic care, research and education, administration and direct service.

The primary goal of societal ethics is not to attend to the unique and specific goals of individuals but to so structure society and allocate resources that the fabric of society in which individuals and institutions exist can be an environment of human flourishing and moral development. But in seeking this common good of society, the good of individuals and the good of orga- nizations cannot simply be ignored. As in the other two realms of beneficence the concern must look in three directions – 3a): primarily to the common good, to the net good of society as a whole, and secondarily to 3b): the good of organizations and 3c): the good of individuals.

Summary of Multi-leveled Beneficence

Mapping ethics as love conceives the heart of human existence as the invitation, opportunity, and duty to care for ourselves and one another in the boundless realm of our minds and hearts (benevolence) and in the limited realm of our actions (beneficence). Because our acts of beneficence are always choices among competing values, goods, claims in three realms – individual, organizational, and societal – we need to construct a system of assistance, a grammar or logic which helps us apply beneficence to the thousands of

unpredictable situations we will meet. This "logic of caring" is the discipline
of ethics. As Maguire says: "To be moral is to love well. How to love well
amid conflicting claims is the problem. Love needs a strategy and that strategy
is ethics" ([22], p. 110).

PART THREE: USING THIS EXPANDED BENEFICENCE MODEL FOR
FURTHER CONCLUSIONS AND EXPLORATIONS

Having roughed out the main elements of an ethic of beneficence I want to
turn to some practical conclusions and explorations. The following discussion is less complete and systematic than I would like. It is a mere beginning
and an invitation to further development.

Given the structure of beneficence as described above there are some key
elements of ethics as "love's strategy" that deserve attention on all three levels.

Community of Concern

The keystone of beneficence is community. The very term beneficence
emphasizes that we are always persons-in-relationships – social beings. It
implies that the natural state of persons is reciprocal, responsive, and engaged.
It understands self-giving as essential and self-realizing; it sees the love
imperative more as an invitation to become than as a constraint on being.

The three-tiered model of beneficence symbolizes how thoroughly individuals are imbedded in layers of social reality: individuals exist within
networks of mediating organizations and these in turn are woven into a matrix
of society. Community is the ocean in which we swim.

In a beneficence ethic there is a presumption that the privileged agent of
ethical discernment is the "community of concern." Certainly, individual ethics
makes sense and *existential ethical decisions* are always made by individuals. But as the three-layered paradigm makes clear at a glance, most ethical
terrain involves community. Beyond this, making *normative ethical judgments* involves the weighing of complex and subtle values and this emerges
primarily from *experience* of these values – from a pool of experience wide
and deep enough to do justice to the issue at hand. Here we are on the wrong
track if professionals' views of patient experience are taken for patient
experience; if men represent the experience of women; if doctors mediate
nurses' views; if administrators speak for the general public. To weigh complex
values we need complex, firsthand experience as well as adequate analysis
of that experience.

Gathering the key elements of this firsthand experience is what the "community of concern" is all about. This is a formal concept to be materially
specified by the issue at hand. The community of concern is constituted by
whatever group is necessary to be in experiential touch with all the major facets
of a beneficence question. Lacking the full community of concern we are in

ethical trouble from the start. No individual or partial group, regardless of ethical fiber or training, can substitute for the full community of concern.

From this perspective, there are two common ethical minefields – one symbolized by hospitals, the other by the Roman Catholic Church. Hospitals are highly structured along lines that stratify, fragment, and compartmentalize. Such a structure is ethically inhibiting, viewed from the perspective of the community of concern. Perhaps the ethics committee movement's greatest contribution can be to introduce a new paradigm of reflection and empowerment into a healthcare structure that necessarily hinders good ethical decision making. The Roman Catholic Church, on the other hand, errs against the community of concern in a different way. By having the same predetermined body of discernment – regardless of the beneficence issue under consideration – this religious group consistently fails to have many key members (different ones for different issues) engaged in the essential community discernment.

An ethic of beneficence will ask early and often: given the nature of the question at hand, do we have the right participants to give us the fullest community of concern possible?

Personal Qualities of Community Members

Beyond a fullness of perspectives on the issue at hand, the members of this community of concern also need to possess some basic personal qualities: a genuine respect for persons; openness to perspectives differing from their own; desire to learn and study; willingness to be wrong; ability to express challenging perspectives directly and honestly; ability to collaborate in a team effort. Ethical wisdom can emerge more consistently and efficiently from a community sharing a heavy dose of these qualities.

Shared Common Vision of Community Members

A key difference between a gathering of special interest advocates and a community of concern is that the latter share a deeper vision that binds them and their differing perspectives into a coherent whole. There may be strong differences on various perspectives of the issue but stronger still are their grounding meanings and priorities. Selecting the community of concern involves finding persons who share this deeper vision or are capable of being called to it. This deeper vision demands attention and resources. It is not simply a given. Elsewhere I have explored how a superficial agreement about justice can hide a deeper level of strong disagreements [15]. A community of concern needs to nurture its shared vision, to test its consensus, to sharpen its definitions, to deepen it, to revise it. Neglect of this deeper vision erodes the community's ability to ethically discern.

Tools of Community Enablement

The community of concern needs to be enabled to do the discernment of beneficence. To harness the complexity of values and disvalues, deeper vision and complementary perspectives, and unity and differences we need instruments of enablement. We can think of these as cognitional tools and process tools.

Cognitional tools: Philosophical and theological ethics can help us understand the importance and role of definitions, distinctions, concepts, principles, and paradigms. These disciplines can provide formal understanding of these elements and material content for application. These disciplines can also suggest methodologies for harnessing this complexity of elements and moving it progressively to closure.

But evaluative knowledge involved in beneficence is more than abstract concepts and cold analysis. Such knowledge is mystic, affective, knowledge of the heart and imagination. Here our resources are not extensively developed and considerable work needs to be done. Fortunately, there is a growing recognition of the direct importance of the arts and literature as tools of enablement for the discerning community. It is in this area that case studies and parables can give human breadth and depth to more discursive principles and definitions.

Group Process Tools: To handle the complex group it has gathered, beneficence needs adequate group process. Adequate process will facilitate a fullness of reflection that:
1. is focused but not rigidly constricting;
2. is co-extensive with the length and breadth of the problem, not ignoring essential areas, not coming to premature closure;
3. attends to persons as well as issues;
4. insures input from all and monopoly of none;
5. allows for self-examination and inter-personal communication;
6. promotes open challenge and confrontation;
7. includes intellection, intuition, affect, and imagination.

Front-end planning of meetings cannot guarantee these characteristics but it can go far in enabling them.

Unity and Diversity of Beneficence Across the Realms

Unity: First, it is important to emphasize the fundamental unity of beneficence/ethics across these spheres. Too often our language masks this instead of emphasizing it. Authors distinguish "morality" from "public policy," the "abstract order of ethics" from the "concrete order of jurisprudence," the "moral order" from "public policy" [4, 9, 11, 17, 28]. Such distinctions have validity. But we need to develop a language that emphasizes the *unity of morality and ethics* as well as the distinctions as we move across these spheres. Otherwise we reinforce the common error that the "real world of ethics" resides in the realm of individual good and that beyond such real ethics lies only the ethical "outback" of politics, common sense, and law. Such misperception blunts our awareness of the most demanding areas of ethics and tends to fragment our moral intellect and imagination. We would benefit from a conceptual model and consistent language that emphasizes the unity of the various dimensions of our ethical life.

Diversity: But beneficence/ethics across these realms is analogous, not univocal, and this involves significant differences between these spheres. A brief sketch of some differences would include the following.

1. All things being equal, as we move from realm 1 to 3, the ethical reality becomes more significant and more complex.

2. Methods, concepts, and principles are presumed not to have the same importance, relevance, and adequacy on one level as they do on another. (For example: The principle of autonomy has an importance on level 1a that it does not sustain on level 2a and is relativized still more on level 3a.)

3. Conclusions reached on one level do not lead to necessary conclusions on another level. (For example: To demonstrate that active euthanasia would be an ethically reasonable option on the individual level does not lead to any significant conclusions on the organizational level and even less so on the societal level.)

4. Substantial deficits on a higher level cannot be corrected by interventions on a lower level. (For example: It is not possible to correct a substantially unjust health care system merely by multiplying the activity of individual hospitals or health professionals.)

5. The ethical character of the higher spheres tends to powerfully define the limits on ethical behavior in the sphere(s) below. (For example: The injustices of a societal system – e.g. Medicaid – will tend to inhibit just behavior of institutions and professionals by punishing those who attempt to behave beyond the boundaries drawn by the system.)

6. Specific disciplines tend to develop awareness/unawareness to different levels of beneficence. (For example: In the United States professional training for social work tends to open awareness to the full range of beneficence more than does professional training for medicine or law.)

7. Different cultures can predispose their members to emphasize one level of beneficence over the others. (For example: According to the statement of Fox/Swazey that for the Chinese "the bedrock and point of departure of medical morality lie in the quality of these human relationships: in how correct, respectful, harmonious, complementary, and reciprocal they are" ([12], p. 650), we would expect this culture to emphasize social beneficence more than individual beneficence. By contrast, the proclivity in U.S. culture is to make the perspective of the individual realm dominant, if not exclusive. This cultural predisposition finds expression in statements such as that by George Annas: "The core legal and ethical principle that underlies all human interactions in medicine is autonomy" ([1], p. 5).

8. Most issues of healthcare ethics have significance on all three levels

but often they are essentially issues of one specific level. It is important to determine which level is dominant and marshal the appropriate instruments for dealing with it on that level.

Developing a fuller understanding of such unity and diversity, as well as their practical implications, constitutes a major challenge for the next phase of healthcare ethics.

Need for Fresh Disciplines

In *The Good Society*, Bellah et al. remark "We need experts and expert opinion, and experts can certainly help us to think about important issues. But democracy is not the rule of experts" ([2], p. 272). This is also true about beneficence – it is not the rule of experts but it needs experts to help think about important issues.

First, it needs a broader band of ethical tradition than has been present up to this time. Fox and Swazey note – and I think legitimately – "It is primarily American analytic philosophy – with its emphasis on theory, methodology, and technique, and its utilitarian, Kantian, and 'contractarian' outlooks – in which most of the philosophers who have entered bioethics were trained" ([12], p. 666). Our paradigm of beneficence suggests that this narrower philosophical tradition is not enough. The latter is most comfortable in the realm of individual beneficence and has a tendency to keep us confined to a "minimalist ethics" [5] rather than push us to the broader bands of beneficence. Political and social philosophy, Christian social teaching, and feminist thought will be especially important for an expanded philosophical horizon.

Our strong proclivity for individual beneficence is reinforced by the extensive presence in bioethics of those trained in U.S. law. Glendon emphasizes that legal training in our society is likely to bring with it a rights-laden ethos and rhetoric which "easily accommodates the economic, the immediate, and the personal dimension of a problem, while it regularly neglects the moral, the long-term, and the social implications" ([16], p. 171).

As we move into the next phase of healthcare ethics we will need the help of such fields as sociology, philosophy/theology of society, anthropology, political science, public health, organizational development, and social psychology. The work of Fox/Swazey and Glendon which sets our own ethical and legal assumptions into a broader social and multi-cultural context point to the potential of such work for an expanded horizon of healthcare ethics. But at present such efforts sit at the margins of medical ethics rather than giving it direction and leadership. I believe that in the next decades there will be a direct relationship between the level of vigor, creativity and fruitfulness of healthcare ethics and the degree to which these now-foreign disciplines assume roles of prominence. Healthcare ethics has benefited from the service of academic ethics and law, but as we move into the wider spheres of beneficence the importance of these disciplines will legitimately wane as the role of other enabling disciplines waxes.

Shaping the Hospital to Beneficence

Rosemary Stevens remarks that the "American hospital has been – and is – a projection of a medical profession whose archetypes are science, daring, and entrepreneurship" ([44], p. 11). Organizations shaped by such archetypes do not promise to be adept at making beneficence decisions in terms we have discussed. If we were to reinvent the hospital to make it a better agent of beneficence what steps might we take?

A first step would be to awaken the institution to the fact that beneficence/ethics is omnipresent – it permeates the life of the organization. Elsewhere I have argued that ethics is coexstensive with human dignity; that wherever decisions are being made about the dignity of persons we have ethical activity. In the conclusion to this article I stated: "Hospital Ethics Committees are just one of many Centers of Ethical Responsibility (CERs) in a modern health care institution; and they are far from the most important. This is so because any group that makes decisions that have impact on the dignity of persons are by definition centers of ethical responsibility. This means that trustees, senior management teams, etc. are essentially 'ethics committees' of great importance – and often even greater importance than the Hospital Ethics Committee. Since all of these CERs are essentially engaged in doing ethics they improve their efforts by attending to key elements of institutional ethics" ([15], p. 286). Helping health care institutions increasingly shape their lives from the perspective of beneficence will involve:

1. helping them recognize the pervasive presence of beneficence/ethics – in planning, budgeting, managing, etc.;
2. helping them identify key components of normative beneficence – community of concern, need for analytic tools and appropriate process, and role of education;
3. helping them build these elements into the fabric of institutional life.

More specifically, considering our model of three-tiered beneficence, I believe that health care institutions will be most myopic concerning their societal agency and responsibility. Reflecting on the history of U.S. hospitals, Stevens observes that without "financial incentives, hospitals have rarely tried to change the system in directions which would clearly be in the public interest." Rather we have a tug-a-war in which "federal and state agencies are trying to further public agendas . . . while hospitals are trying to deflect and/or resist them, in order to further their own organizational objectives" ([44], p. 353).

Hospitals *need to develop a conscience concerning sphere 2b of beneficence* – a moral sensitivity to their societal role and their responsibility for the healthcare common good. They need to become a more vigorous part of the community of concern shaping our health care system and less an agent of special interest pressure. They need to invest time in *understanding health issues from the perspective of community good, not merely from the viewpoint of organizational benefit*. They need to take time to develop some consensus about what an adequate health system for our society would look like; what percentage of the GNP we can reasonably spend on health care – relative to the other needs of the common good. If the health care professionals of our nation do not have a coherent set of questions and answers around these central issues we should not be surprised if politicians reduce such issues to matters of balancing a budget. However one judges the substance of the plan proposed by Physicians for a National Health Program, their substantial efforts at advocacy for the common good present us with an outstanding – and almost unique – example of "2b conscience" in the world of healthcare organizations and professionals.

In approaching these issues the larger hospital community can benefit from some findings of ethics committees. Two findings are of special importance. First, education is an essential element for good ethical decision making. Balancing and weighing complex and subtle dimensions of human, inter-dependent good requires an informed community. Devising an appropriate public policy on quality, access and financing of health care requires a community of committed and informed discussants. If hospitals are to move toward an ethic of societal interest they will have to engage their governance and management in ongoing education about such issues. Good institutional ethics implies a commitment to substantive education for trustees and management teams.

Beyond this lies the second and even more fundamental question of the community of concern. To what extent does a typical hospital board represent the community of concern one would want for making the range of beneficence decisions sketched above? If it is lacking, how should we enhance this community in order to give it more the character of an appropriate community of concern? For example, two perspectives that are commonly absent from governance that seem essential are – internally – the nursing perspective and – externally – the public health perspective.

Questions of selection of leadership – in terms of specific criteria that enjoy consensus among the selection group, and effective ways of using these criteria to screen candidates – has not received much ethical attention. Given the central role of the community of concern in making good ethical decisions, hospitals should give considerably more attention to the substance and process of leadership selection [13, 14].

New Phase, New Tasks for Healthcare Ethics

I believe that phase one of healthcare ethics is coming to a close. This first phase has spanned the last twenty years and generated literature, processes, and tools that have primarily focused on individual ethics. Given our cultural fixation on individual rights and the paternalistic ethos and mores of U.S. medical care in those years, such an emphasis is historically understandable. But there are increasing signs that the limits of such emphasis are being experienced and the need for attention to the broader realms of beneficence is being recognized.

In phase two an unaccustomed set of questions arise at the very start of ethical examination of an issue: what significance does this issue have on the three levels of beneficence? which level(s) need we deal with? is one level *the decisive level of concern*? Only when we have answered these questions adequately can we effectively do ethics – by gathering the appropriate community of concern and providing the level-appropriate tools of enablement to the community.

For instance, to miss the decisive level of ethical relevance is to miss the ethics of the issue whole and entire. As Callahan says about devising an ethically sound system of healthcare: "A society that thinks of illness as simply an individual phenomenon, with an occasional public face, is already on the wrong track. . . . We have no lasting hope of devising a decent understanding of health – and thus of fashioning a viable healthcare system – unless we learn better how to attend to the social dimension of health, indeed unless we learn how to shift our priorities sharply in a societal direction" ([6], pp. 104–105).

Daniel Maguire expresses the same concern about the ethical issue of affirmative action and the Supreme Court's handling of that question. "Affirmative action is an issue of social-distributive justice. The Court usually attempts to handle it using the concepts appropriate for individual (commutative) justice. It doesn't work" ([24], p. vii).

In the terms of this essay, Callahan and Maguire are saying that healthcare policy and affirmative action are primarily and essentially issues of societal beneficence and cannot be grasped or resolved in terms appropriate for individual beneficence. In fact, trying to handle common good with categories of individual good represents ethical disaster on two counts: first, we totally miss the ethical substance of the issue, but worse, we are fooled about this because we've engaged in some ethical discussion.

I believe that this error of casting societal issues in individual terms is a major factor in our abortion turmoil. We run the risk of repeating this methodological error with other issues of societal beneficence such as surrogate motherhood, fetal tissue experimentation, and euthanasia. In fact, medical ethics will have to reckon with this chronic illness of collapsing the social into the individual for some time to come. Lacking a congenial culture for common good consciousness and few tools of community enablement for this realm, our natural response to ethical reality is to snatch up the arms of individual beneficence and rush into battle. A preventive treatment for such reductionist rashness will be to make our first ethical move one of mapping the relevance of a given issue on the three levels of beneficence and determining which level represents the ethical center of gravity.

This leads to another high priority item in phase two of healthcare ethics – *the development of ethical tools for organizational and societal ethics, analogous to the ones already generated for individual ethics*. The volume and complexity of this work will be exponentially greater than that of phase one.

We will need to develop an integrating conceptual model that honors the unity and diversities across these realms of ethical complexity (adopting, adapting or replacing altogether my proposed three-sphere model). We will need concepts, definitions, distinctions, principles, paradigms and method-ologies proportionate and appropriate to the varying scope and complexity of social beneficence, whether organizational or societal.

A brief look at Daniel Callahan's *What Kind of Life* and Daniel Maguire's *A Case for Affirmative Action* can illustrate the kind of work this involves. I suggest viewing them as models of the kind of invention and creation that needs doing.

In Callahan, for example, instead of the cornerstone of individual benefi-cence – autonomy – we find such cardinal principles as "limits." "The dominant future policy bias in our system, I contend, should be that of cultivating a sense of boundaries, finding ways of dampening our unbounded hopes and enthu-siasms" ([6], p. 161). Another cardinal preference principle is "the principle of health symmetry." Those options are to be preferred which offer the "likelihood of enhancing a good balance between the extension and saving of life and the quality of life" ([6], p. 164). He suggests the "principle of technological success": "A technology that might achieve a short-term medical success but may generate long-term medical or social problems requires special scrutiny and considerable resistance" ([6], p. 164). This would have us examine the "medical consequences of success," the "societal implications of success," and the "cultural impact of success" ([6], pp. 164–171).

Maguire's work develops some distinctions important for handling affir-mative action as an issue of societal beneficence. For example he discusses at length the content of and differences between individual good and common good ([24], pp. 93–108). He also develops a more complex and fuller concept of justice than is usual in the current discussion [23, 24]. "Quite simply there are three forms of justice because persons relate to persons in three different ways. We relate on a one-to-one basis (individual justice); the individual relates to the social whole (social justice); and the representatives of the social whole relate to individuals (distributive)" ([24], p. 73). To facilitate the application of his principle of "redistribution" (an alien ethical category for one dedi-cated to individual beneficence) he develops four criteria for preferential aid to disempowered groups:
1. No alternatives to enforced preference are available.
2. The prejudice against the group must reach the level of depersonaliza-tion.
3. The bias against the group is not private or narrowly localized but is rather entrenched in the culture and distributive systems of the society.
4. The members of the victim group must be visible as such and thus lack an avenue of escape from their disempowered status ([24], pp. 141–142).

Even in this quick glance one can get an idea of what I mean by "inventing the tools and methodologies of social ethics." When we think of the time and energy that it cost to develop a consensus around the tools of 1a benef-

etc. – we know that building an analogous body of consensual tools for the spheres of organizational and societal ethics represents a monumental task. I believe that the single most important challenge of the next era of health-care ethics will be the development of a community consciousness of beneficence beyond the individual and crafting the tools of enablement for these broader realms.

Codes of Ethical Wholeness

This model of beneficence would urge us to formulate our codes of ethics to cover all the realms of ethics. One notable example of such a comprehensive professional code that could serve as a model is the National Association of Social Workers' Code of Ethics [20]. This code identifies six areas of concern:
1. personal conduct;
2. duties to clients;
3. duties to colleagues;
4. duties to employers;
5. duties to the Social Work Profession;
6. duties to Society.
Not only does this code recognize this comprehensive range of concerns, it also translates each area into an effective and realistic level of specificity.

Even some of the recent attempts to develop richer professional medical codes stay within the realm of individual ethics ([29], pp. 203–206). Certainly physicians owe priority to their patients. Their codes should reflect this. But these codes should also speak at some level of specificity about duties on the organizational level – to hospitals, to nurses, to the medical staff, as well as on the societal level – to help develop and maintain a health care system with just patterns of access and reasonable parameters of quality and cost.[1]

Ethics committees could revisit the issue of their own mission/purpose statement in light of these three realms of ethical complexity. Are there some areas that deserve more explicit attention? If so, would this involve a change/expansion of membership? Does this modify the traditional big three activities of the committee: consultation, education, policy review and development? To what extent should IECs become agents of social change within their own institutions and within the larger community?

Hospitals could develop a code of ethics for their institution. Ethics committees could help by developing a draft of such a code. It would give priority to organizational good but explicitly set this within the complexity of individual and societal good. Such an exercise would contribute to the self-education of the ethics committee and offer an opportunity to engage in the ethical education of both governance and management. Such a code of ethics could involve the medical staff and employees of the hospital in its evolution.

As a nation we do not have a "National Code of Health Care Ethics." One

As a nation we do not have a "National Code of Health Care Ethics." One modest but real contribution to the realization of an ethically sound national health program could be a broad effort to formulate such a "National Code." This would press us to engage in the troublesome but indispensable discussion of key elements of the health care common good instead of merely debating concrete proposals without a larger frame of reference. It would also force us to struggle with the interdependencies and shifting hierarchies that prevail between these three spheres of beneficence.

Beneficence ethics invites us to evolve codes of ethics that explicitly face the conflicting loyalties that confront us within each sphere and across the spheres of the comprehensive picture of beneficence.

Global Beneficence

There is a fourth sphere of beneficence about which I have been silent up to now – the sphere of global beneficence. The concern here is the common good of the globe. It represents a still broader, more complex and vastly more difficult realm of values/goods/loyalties in conflict. I mention it here only to sketch the complete parameters of a beneficence ethic, not to fill in the details. However, two perspectives deserve passing mention. First, global beneficence considers the good of the planet itself as an ecology and environment for the human family. It demands that we think in space- and time-frames that are alien and alienating to our current patterns. Our concern here is the kind of health environment we are shaping globally for the generations to come. Second, global beneficence looks at the balanced allocation of healthcare resources throughout the global community. While allowing for differences between nations it explores the limits to those differences and the responsibilities of wealthy nations for the health care of poorer nations. An ethic of beneficence not only presses us to recognize this further realm but it also offers a framework – analogous to the other spheres – for filling in the specifics.

Summary

Drawing on the Christian tradition of beneficence as the heart of morality and ethics offers us the chance to view the enterprise of healthcare ethics from a fresh perspective. It highlights the constant process of resolving our conflicting loyalties on three levels of human reality – the individual, the organizational, and the societal. It invites us to develop a consciousness of new ethical horizons and to create the ethical tools needed by the community to meet the challenges of these new realms.

Center for Healthcare Ethics
St. Joseph Health System
Orange, California
U.S.A.

NOTE

[1] The St. Joseph Health System in Orange, California has promoted the development of physician value statements among the medical staffs of its hospitals. These "ethical codes" take some meaningful steps in including the broader ranges of beneficence.

BIBLIOGRAPHY

1. Annas, G.: 1990, 'Life, Liberty, and Death', *Health Management Quarterly*, vol. 12, No. 1, 5–8.
2. Bellah, R., Madsen, R., Sullivan, W., Swidler, A., and Tipton, S.: 1991, *The Good Society*, Alfred A. Knopf, New York.
3. Bellah, R., and Sullivan, W.: 1987, 'The Professions and the Common Good: Vocation/Profession/Career', *Religion and Intellectual Life* (Spring), vol. IV, no. 3, 7–20.
4. Bouchard, C., and Pollock, J.: 1989, 'Condoms and the Common Good', *Second Opinion* 12, 98–106.
5. Callahan, D.: 1981, 'Minimalist Ethics', *Hastings Center Report*, vol. 11, no. 5 (October), 19–25.
6. Callahan, D.: 1990, *What Kind of Life*, Simon and Schuster, New York.
7. Carpentier, R.: 1953, 'Vers une morale de la charité', *Gregorianum* 34, 47–62.
8. Childress, J.: 1978, 'Just War Theories: The Bases, Interrelations, Priorities, and Functions of Their Criteria', *Theological Studies*, vol. 39, no. 3, 427–445.
9. Curran, C.: 1987, *Toward an American Catholic Moral Theology*, Notre Dame University Press, Notre Dame, Indiana.
10. Dewey, J.: 1965, 'The Nature of Moral Theory', in R. Ekman (ed.), *Readings in the Problems of Ethics*, Charles Scribner's Sons, New York, pp. 4–8.
11. Farrell, W.: 1990, 'A Note on the Abortion Debate', *America* (January) 27, 52–53.
12. Fox, R., and Swazey, J.: 1988, 'Medical Morality is not Bioethics – Medical Ethics in China and the United States', *Essays in Medical Sociology*, Transaction Books, New Brunswick, 645–671.
13. Glaser, J.: 1989, 'Selecting the Cream of the Crop', *Health Progress* (July–August) 86–89.
14. Glaser, J.: 1992, 'Selecting the Cream of the Crop (Part II)', *Health Progress* (April) 14–16, 33.
15. Glaser, J.: 1989, 'Hospital Ethics Committees: One of Many Centers of Responsibility', *Theoretical Medicine* 10, 275–288.
16. Glendon, M.: 1991, *Rights Talk*, The Free Press, Macmillan, New York.
17. Griffin, L.: 1990, 'The Church, Morality, and Public Policy', in C. Curran A (ed.), *Moral Theology: Challenges for the Future*, Paulist Press, New York, pp. 334–354.
18. Gustafson, J.: 1990, 'Moral Discourse About Medicine: A Variety of Forms', *The Journal of Medicine and Philosophy* 15, 125–142.
19. Hardin, G.: 1968, 'The Tragedy of the Commons', *Science* 162, 1243–1248.
20. Hepworth, D., and Larsen, J.: 1990, *Direct Social Work Practice: Theory and Skills*, Wadsworth, Belmont, CA.
21. Jennings, B., Callahan, D., Wolf, S., et al.: 1987, 'The Public Duties of the Professions', *Hastings Center Report*, vol. 17, no. 1, 1–20.
22. Maguire, D.: 1979, *The Moral Choice*, Winston Press, Minneapolis.
23. Maguire, D.: 1983, 'The Primacy of Justice in Moral Theology', *Horizons* 10: 72–85.
24. Maguire, D.: 1992, *A Case for Affirmative Action*, Shepherd Inc. Dubuque.
25. McCormick, R.: 1978, 'Notes on Moral Theology', *Theological Studies*, vol. 39, no. 2., 76–136.
26. McCormick, R.: 1978, *Doing Evil to Achieve Good*, Loyola University, Chicago.
27. McCormick, R.: 1981, 'Notes on Moral Theology', *Theological Studies*, vol. 42, no. 1, 74–121.

28. McCormick, R.: 1983, 'Bioethics in the Public Forum', *Milbank Memorial Fund Quarterly/Health and Society*, vol. 61, no. 1, 113–126.
29. Pellegrino, E., and Thomasma, D.: 1988, *For the Patient's Good: The Restoration of Beneficence in Health Care*, Oxford University Press, New York.
30. Pope Pius XI.: 1931, 'Quadragesimo Anno', in C. Carlen (ed.), *The Papal Encyclicals: 1903–1939*, Pieran Press, Raleigh.
31. Rahner, K.: 1965, 'Über die Einheit von Nächsten- und Gottesliebe', *Schriften Zur Theologie, VI*, Benziger Verlag, Einsiedeln, 277–300.
32. Rahner, K.: 1983, *The Love of Jesus and the Love of Neighbor*, Crossroads, New York.
33. Schreiner, J.: 1966, *Die Zehn Gebote im Leben des Gottesvolkes*, Patmos, Düsseldorf.
34. Schüller, B.: 1966, 'Zur theologischen Diskussion über die lex naturalis', *Theologie und Philosophie* 4, 481–503.
35. Schüller, B.: 1970, 'Typen ethischer Argumentation in der katholischen Moraltheologie', *Theologie und Philosophie* 4, 526–550.
36. Schüller, B.: 1970, 'Zur Problematik allgemein verbindlicher ethischer Grundsätze', *Theologie und Philosophie* 1, 1–23.
37. Schüller, B.: 1971, 'Zur Rede von der Radikalen Sittlichen Forderung', *Theologie und Philosophie* 3, 321–341.
38. Schüller, B.: 1972, 'Direkte Tötung – indirekte Tötung', *Theologie und Philosophie* 3, 341–357.
39. Schüller, B.: 1974, 'Neuere Beiträge zum Thema Begründung sittlicher Normen', *Theologische Berichte* 4, 109–181.
40. Schüller, B.: 1976, 'Typen der Begründung Sittlicher Normen', *Concilium*, 648–654.
41. Schüller, B.: 1977, 'Die Bedeutung der Erfahrung für die Rechtfertigung sittlicher Verhaltensregeln', *Christlich glauben und handeln: Fragen einer fundamentalen Moraltheologie in der Diskussion*, Patmos, Düsseldorf, 261–286.
42. Schüller, B.: 1980, *Die Begründung Sittlicher Urteile*, 2nd ed., Patmos Verlag, Düsseldorf.
43. Stark, W.: 1958, *The Sociology of Knowledge*, London.
44. Stevens, R.: 1989, *In Sickness and in Wealth, American Hospitals in the Twentieth Century*, Basic Books, New York.
45. United States Catholic Conference Administrative Board, 1984, 'Political Responsibility: Choices for the 80's', *Origins* 13, April 12, 732–736.
46. United States Catholic Conference, 1986, *Economic Justice for All*, Washington, D.C.
47. Veatch, R.: 1977, *Case Studies in Medical Ethics*, Harvard, Cambridge.

KAREN LEBACQZ

EMPOWERMENT IN THE CLINICAL SETTING

In *Professional Ethics: Power and Paradox* [17], I proposed that the power of the professional person is morally relevant to determining what should be done in the practice setting and that justice or empowerment of the client becomes a central norm for professional practice.[1] The time has now come to see what justice as empowerment means for the clinical setting. The task is particularly crucial in light of Kapp's recent definition of empowerment as advocating for oneself and participating maximally in one's own significant decisions ([14], p. 5). Under this definition, to choose dependence upon others is to "forego" empowerment ([14], p. 6). I will explore below the adequacy of this definition.

Just as we learn about justice by exploring experiences of injustice [16], so we may learn about empowerment by exploring disempowerment. Disempowerment, and therefore empowerment, within the clinical setting will differ from setting to setting and from population to population. I will examine two populations: those who begin with power and become disempowered in the clinical setting, and those who begin from a position of relative powerlessness and experience the clinical setting from that perspective.

THE DISEMPOWERMENT OF THE POWERFUL

Hans Jonas once argued that the ideal research subject is a doctor: the doctor is best positioned to understand the risks and implications of the research, and to give a truly voluntary and informed consent [13]. In short, the doctor as subject is most likely to be empowered in the research setting. Similarly, the doctor as patient is most likely to be empowered in the clinical setting. By looking at the disempowerment experienced by those most likely to retain power in the clinical setting, we begin to develop a sense of what constitutes disempowerment and therefore what would constitute empowerment. So we begin with those who experience no language, cultural, knowledge, or sexual barriers to empowerment.

Yet even white, male, well-educated doctors, when they become patients, experience disempowerment in the clinical setting. The film "The Doctor," based on *A Taste of My Own Medicine* by Ed Rosenbaum [22] and just released at the time of writing this essay, demonstrates precisely this disempowerment. Here is a physician, a surgeon, accustomed to ordering people around in the hospital who must now wait in line, fill out forms by the hour, be told

133

G.P. McKenny and J.R. Sande (eds.), Theological Analyses of the Clinical Encounter, 133–147.
© 1994 *Kluwer Academic Publishers. Printed in the Netherlands.*

"sorry, the doctor is late today," and undergo any number of forms of indignity experienced routinely by patients.

A similar story is told in Oliver Sacks' delightful treatise *A Leg To Stand On* [24]. Sacks broke his leg in a climbing accident. Not only was the leg broken; Sacks also lost all sense of feeling in the leg, and all ability to move or exercise voluntary control over it. He lost "proprioception," the sense of owning one's own limbs and having command over them. Ironically, proprioception is one of the foci of Sack's work as a neurologist. Thus, his story is the story of a man who moves not only from the status of physician to the status of patient, but indeed, to the very kind of patient that he himself treats. He thus learned to see from the "inside" what his patients had tried to communicate to him.

Two Miseries, Two Empowerments

To be a patient – at least under critical circumstances – is to live in an altered world. It is almost as though one has entered the "twilight zone." Perceptions are distorted, time and space appear different, even everyday conversation can loom threateningly: "'Execution tomorrow,' said the clerk in Admissions. I knew it must have been, 'Operation tomorrow,' but the feeling of execution overwhelmed what he said" ([24], p. 46).

Illness takes place on two levels. Sacks calls it "two miseries." One is the physical disability, the "organically determined erosion of being and space" ([24], p. 158). The other he calls the "moral" dimension associated with "the reduced stationless status of a patient, and, in particular, conflict with and surrender to 'them' – 'them' being the surgeon, the whole system, the institution. . . ." ([24], p. 158).

Illness is not just a matter of physical disability. The clinical setting involves also, and even more importantly, a change in one's structural and sociological status. One becomes a "patient."

Even the most powerful of patients therefore feel disempowered in two ways. First, they have lost some ability previously possessed – in Sacks' case, the ability to walk and even to feel his own leg. Loss of ability is annoying at best, frightening at worst: "I found myself . . . scared and confounded to the roots of my being" ([24], p. 79). This is the first "misery" with which the patient deals, and it is often overwhelming in itself, undermining one's ability to deal normally with the world.

But patients have also lost normal status in the world. For Sacks, and for many others, it is the role of patient as much as physical disability itself that inflicts misery and requires empowerment. "I felt morally helpless, paralyzed, contracted, confined – and not just contracted, but contorted as well, into roles and postures of abjection" ([24], p. 158). For example, Sacks asked for spinal rather than general anesthesia, and his request was denied. He writes, "I felt curiously helpless . . . and I thought, Is *this* what 'being a patient' means?" Sociologists have described the "sick role" or status of patient. From the inside, this status is often a feeling of diminishment. Sacks calls himself a "man reduced, and dependent" ([24], p. 133).

The role of patient is reinforced by institutional structures and practices: "we were set apart, we patients in white nightgowns, and avoided clearly, though unconsciously, like lepers" ([24], p. 163). As Alan Goldman puts it in his examination of professional ethics,

life in hospitals . . . continues to be filled with needless rituals suggestive of patient passivity, dependence, and impotence. The institutional setting is still structured in such a way as to block the exercise of rights at least partially accepted intellectually. . . . Patients are rarely permitted to see their charts; pills are almost literally shoved into their mouths. . . . Often newly admitted patients perfectly capable of walking are taken to their rooms in wheelchairs, an apt symbol of the helpless pose they are made to assume from the time of their entrance into this alien and authoritarian setting ([10], pp. 224–5).

Ultimately, suggests Sacks, he indeed became an "invalid": in-valid ([24], p. 164). Sacks resisted the patient role at every step. Yet he recognized that both he and the surgeon were, in a sense, "forced to play roles – he the role of the All-knowing Specialist, I the role of the Know-nothing Patient" ([24], p. 105).

If illness is composed of two miseries, then recovery will require two empowerments. "Now we needed a double recovery – a physical recovery, and a spiritual movement *to* health" ([24], p. 164). The patient needs physical healing – in Sacks' case, surgery and physical therapy so that he could once again walk, run, jump, and do things that his body had "forgotten" how to do. But as much as the physical healing, the patient needs recovery from the abject, reduced, dependent status of patient.

Both miseries are disempowering. But it is the "contortion" into roles and postures of abjection that is the core of the power gap between physician and patient. Such contortion need not accompany physical deterioration. Different structures, a different approach in the clinical setting, might significantly reduce the second "misery."

Nor is such contortion necessarily diminished when physical healing takes place. Too often, we assume that once health is restored, the patient automatically becomes a non-patient and experiences restored moral status. In my experience, this is not true. Effects of dependent status can linger, making future contacts difficult and undermining patients' sense of their own worth and being. While the patient may literally move outside the hospital or clinic and cease being a patient in a technical sense, the psycho-sociological effects of dependent patienthood may remain. Moreover, during the time of clinical care, the dependent status of patient can adversely affect medical treatment.

What can be done to empower patients? The loss of function, the physical disability, is the initial presenting problem. The best the medical team can do in the face of it is to try to heal. But is the loss of status, the diminished sense of personhood that often accompanies being a patient, also necessary? Must there be two miseries? Is there a way to reduce the second misery, to hasten recovery from it, and to empower the patient who experiences it? Using

Sacks' experience, we can examine the disempowerment of the powerful and suggest how different structures and responses might empower the patient.

The Central Role of Communication

The clinical context begins with communication. Disempowerment begins with failures of communication. One of the most disempowering things that happens in the clinical context is shutting down the patient's words.

In Sacks' case, this began with his first attempts to share what had happened. "They wanted to know the 'salient facts' and I wanted to tell them everything – the entire story" ([24], p. 47). Something had *happened* to Sacks. It was his story. He wanted to tell it. And he wanted to be heard. So from the first "intake" interview in which Sacks tried to tell the "entire" story and the medical team asked for the "salient facts," things went awry.

Failure to communicate went far beyond this first incident. Sacks knew that something was wrong with his leg because he had lost all feeling in it. He waited (and waited and waited!) for the surgeon to come in order to raise his concerns. The surgeon finally appeared, only to state briskly, "there's nothing to worry about" and disappear before Sacks could say more than "but. . . ." Sacks was given no *time* to communicate.

As the days went by and the leg failed to respond to physiotherapy, Sacks became desperate to communicate his concern: "Desperately now, I wanted communication, and reassurance" ([24], p. 88). Above all, he recognized a need to communicate to the surgeon and have the surgeon understand. While he wanted reassurance, he was prepared to accept the truth if no reassurance could be given: "I should respect whatever he said so long as it was frank and showed respect for me, for my dignity as a man" ([24], p. 93).

When the surgeon came, he neither looked at Sacks nor spoke to him, but turned to the nurse and said, "Well, Sister, and how is the patient now?" ([24], p. 104). Rather than respect and frankness, Sacks was treated as a nonentity. He was not even addressed by the surgeon, but talked about as if he were not there. Sacks was given no *respect* for himself as a communicator; all the communication was with those around him. Using Kapp's definition of empowerment as advocating for oneself and participating "maximally" in one's own significant decisions, Sacks was clearly disempowered.

Sacks persisted in raising his concern and tried, falteringly, to tell the physician what was wrong:

It's . . . it's . . . I don't seem to be able to contract the quadriceps . . . and, er . . . the muscle doesn't seem to have any tone. And . . . and . . . I have difficulty locating the position of the leg ([24], p. 104).

I have quoted this speech as Sacks describes it. If it is an accurate representation of what he said, this fact alone is significant. Sacks is a literary man. He writes eloquently, powerfully.[2] It is difficult to imagine him at a loss for words, or stumbling over his words. Yet, confronted with the power of the physician, and in his own dependent state as patient, stumble is apparently what he did. He seems to have stammered, acted hesitant and evidenced confusion.

There may be an important lesson here for empowerment in the clinical context. Few medical people realize how dis-empowering the very context is. Patients generally feel inadequate in their descriptions of what is wrong. They hesitate, stumble, try to find the right words. Nothing seems to come out right. The patient who stumbles over her words is not necessarily stupid, but may simply be experiencing, as Sacks did, a diminishment of her capacity to verbalize.

I went to my physician complaining of pain in my hip joint. He asked me to stand and turn in certain ways, and then declared flatly that it could not be my *joint* which was hurting. It must be the *tendon*, not the joint. To him, this technicality and diagnostic accuracy is very important. To me, only the pain that makes it hard for me to climb stairs is important. I do not care whether the pain originates in the joint, technically speaking, or in the tendon. I care only about what can be done to alleviate it, since I live in a house full of stairs. But his focus on the technicalities made it difficult for me to persist in my query. I had been told that I was *wrong*. I felt inadequate and unable to communicate. I gave up, and no treatment was forthcoming.

Sacks suggests that there is among doctors, "in acute hospitals at least, a presumption of stupidity in their patients" ([24], p. 171). Whether all doctors do in fact consider their patients stupid, in acute or other contexts, failures to communicate often have the subtle effect of giving the patient a sense that she or he is not only stupid but also not worthy of the physician's time and effort.

Failure to hear the patient's story, impatience to get to salient facts, lack of time to listen, failure to address the patient at all, focusing on technicalities or calling the patient's understanding wrong – all these are disempowering in the clinical context. The patient who has been treated this way often gives up on the effort to communicate.

Many feel keenly their "ex-communication" ([24], p. 110). Not only have they been shut out from the healthy world literally – stuck in the hospital, wrapped in white gowns, and avoided by healthy people – but they are now shut out symbolically by failure to communicate, to listen, to honor their perspective. "As a patient in the hospital I felt both anguish and asphyxia – the anguish of being confronted with dissolution, and asphyxia because I would not be heard" ([24], p. 209). Thus does Sacks describe the life-killing effect of having communication shut off.

Lack of communication is not the only thing that is disempowering in the clinical context. Being denied a legitimate request (e.g. for spinal rather than general anesthesia), being forced to wear unattractive hospital gowns that strip one's individuality, being shunted from department to department like a sack of potatoes – all these and many other routine aspects of clinical care also take power away from patients. But many of these ills would be compensated by careful communication that leaves patients feeling as though they have been treated as persons, as though they can advocate for themselves and participate maximally in decisions. As Sacks puts it, he would have been content with whatever he was told, so long as it was told with respect.

The Liberating Word

If failures of communication are the beginning of disempowerment, then communication can be the beginning of empowerment: "The posture, the passivity of the patient, lasts as long as the doctor orders. . . ." ([24], p. 133). Sacks points to the importance of the liberating word on several occasions. In order to heal, to regain use of his leg, he had to walk. Rehabilitation is based on action. Yet that action was birthed not just by himself, but by others: "I had to *do* it, give birth to the New Act, but others were needed to deliver me, and *say*, 'Do it!'" ([24], p. 182). He calls this speaking the essential role of the teacher or therapist. It is a form of midwifery. Only as others granted permission could he find the way to do something new.

Once he missed a memorial service that he would have liked to attend, and lamented to the nurse that he was unable to go. "Why not?" she queried. By challenging the limitations he had set for himself, she removed them: "The moment she spoke and said, 'Why not?' a great barrier disappeared. . . . Whatever it was, I was liberated by her words" ([24], p. 184).

Words of support ("do it") and words of challenge ("why can't you?") can both be liberating. Both can set the patient free to take a new step in the healing process and to claim skills and territory that the patient has not been able to claim by herself.

Empathy

But communication goes far beyond words. When Sacks was first injured, a young surgeon danced into his room, and leaped on the bed-side table. This surgeon had once had a broken leg, and showed Sacks the scars from surgery. "He didn't talk like a text-book. He scarcely talked at all – he acted. He leapt and danced and showed me his wounds, showing me at the same time his perfect recovery" ([24], p. 44). This visit made Sacks feel "immeasurably better." Here, he encountered someone who had been through it and could demonstrate that there is light at the end of the tunnel. Later, another surgeon came to see Sacks, and Sacks felt that he could communicate with this man. "*I've been through this myself*," said the surgeon, "I had a broken leg. . . . *I know what it's like*" ([24], p. 183). The empathy that comes from experience communicates and empowers.

Empathy gives authority to speak: "So when Mr. Amundsen said that the time had come to graduate, and give up one crutch, he spoke with authority – the only real authority, that of experience and understanding" ([24], p. 183). Sacks reflects on his own change as a physician because of what he went through as a patient, "Now I *knew*, for I had experienced myself. And now I could truly begin to understand my patients" ([24], p. 202). At the end of the film "The Doctor," the protagonist puts all his physicians-in-training through the experience of being a patient in the hospital. Sacks suggests that there is an "absolute and categorical difference" between a doctor who knows and one who does not, and that this difference is because of the personal experience of "descending to the very depths of disease and dissolution" ([24], p. 203).

There is here, then, an important epistemological question that relates to empowerment. One who knows what the patient suffers and can truly hear the patient can empower the patient. But how does one "know" what the patient suffers? Those who have been through a similar experience have readiest access to empathy. This suggests that, where possible, medical care teams should include at least one care-giver who has experienced what the patient suffers. Where this is not possible, groups of patients with similar problems might be assembled. Patients often feel more secure about their position and more enabled to question medical practice when they are with a group that shares their experience.

Art and Religion

Sacks found several other things empowering as well. When he was rebuffed by the surgeon who told him that his concerns were "nothing," he felt as though he had entered a scotoma, "a hole in reality itself" ([24], p. 109). "In this limbo, this dark night," he writes, "I could not turn to science. Faced with a reality, which reason could not solve, I turned to art and religion for comfort. . . ." ([24], p. 114). Two additional sources of empowerment, then, are art and religion.

A friend loaned Sacks a tape recorder with only one tape: Mendelssohn's violin concerto. "Something happened" to Sacks from the first playing of the music. The music appeared to reveal the creative and animating principle of the world, and thus began to give him hope for animation of his own portion of the world, his leg. "The sense of hopelessness, of interminable darkness, lifted" ([24], p. 119). When he first tried to walk on crutches, he was unable to do so until suddenly the music began to play in his mind, and then he found that he could move to the rhythm.

In his own medical practice, Sacks finds that music can "center" his patients. It appears to restore a sense of the inner self that has been lost through neurological injury or disease ([24], p. 219). Because healing and empowerment involve both physical rhythms of the body and also the "center" of the self, music might be a powerful tool for empowerment. "Music," writes Sacks, "was a divine message and messenger of life. It was quintessentially quick – the 'quickening art,' as Kant has called it. . . ." ([24], p. 148). This makes me wonder what would happen if our hospitals provided not television sets but stereos equipped with the great masterpieces of music from the centuries!

Art is not the only response to realities which reason cannot solve. "Science and reason could not talk of 'nothingness,' of 'hell,' of 'limbo'; or of 'spiritual night.' They had no place for 'absence, darkness, death.' Yet these were the overwhelming realities of this time" ([24], p. 114). In order to find a language adequate to describe his experiences, Sacks turned to religion. The patient who faces dissolution of her world needs a language adequate to give voice to that dissolution and to provide a framework within which it can be understood, accepted, and overcome. The language of science and reason is often too sterile for this task. The language of religion, precisely because it is often poetic [5] and mysterious, is adequate to the task.

Sacks gives eloquent expression to the power of religious language when he writes, "In a sense my experience had been a religious one – I had certainly thought of the leg as exiled, God-forsaken, when it was 'lost' and, when it was restored, restored in a transcendental way" ([24], p. 190). While he admits that his experience was also a "riveting scientific and cognitive" experience, it had transcended the limits of science and cognition. A language beyond science was needed.

Moreover, it is not only the *language* of religion that is empowering. Sacks went home for a night to see his family. While there, he attended synagogue. Here he experienced "inexpressible joy": "Behind my family I felt embraced by a community and, behind this, by the beauty of old traditions, and, behind this, by the ultimate, eternal joy of the law" ([24], p. 189). Religion is not just a language. It is a community, a set of laws and rituals, a sense of belonging to something larger and more grounding than one's own family or personal universe. All of these things have empowering possibility. They also suggest that Kapp's definition of empowerment is too individualistic and based too much on an autonomy model. Empowerment includes community and connection; it includes strengthening and honoring relationships.

Summary

What Sacks needed was "a leg to stand on." He needed it in two senses: the literal, physical healing of his limb, and the symbolic, "moral" healing of his status in the world. Because there were two miseries, two empowerments – two "legs" – were needed. The second leg is social and spiritual. Although the empowering possibilities of physical healing should not be underestimated, neither should the need for the second leg be neglected. It has to do with the meaning system of the patient, with hope and fear, with anxiety and joy, with community and solidarity. It is a leg composed of the liberating word, the communication that comes from empathy, the centering power of music and art, the adequacy of religious language and the solidity of religious ritual and community.

Oliver Sacks was one of the lucky ones. White, male, well-educated, a physician to boot, he was in a position to be as powerful as any patient can be. The mere fact that one so powerful experienced two "miseries" and needed two empowerments gives us many clues as to what happens in the clinical setting. But it does not cover the situation of those who are not powerful at the outset. What about those who suffer language, educational, racial, or sexual barriers when confronting the medical establishment?[3] The experiences of patients who are female, non-white, poor, not well educated, or in some other way less powerful than the white male physician suggest some additional dimensions of disempowerment and therefore of empowerment. Those who begin in a more powerless position have many more barriers to empowerment. For them, it will take not only a change in communication, a bit more thoughtfulness, or a little music to give them back their moral status and sense of wholeness. For them, it will take nothing short of a change in the system.

Consent and Rationality

Consider, for example, the case of Maria Diaz, whose doctors recommended tubal ligation while she was in the last stages of a difficult labor. "I told them I would not accept that. I kept saying no and the doctors kept telling me that this was for my own good" ([7], p. 108). Maria Diaz never agreed to be sterilized and signed no consent forms for the procedure; but she was sterilized during caesarian section. Later, she and other Hispanic women brought a suit against U.S.C.-L.A. medical center where the procedure was performed. In a subsequent study, Dreifus and her colleagues found that nine of 23 physicians interviewed had either witnessed coercion or worked under conditions that border on coercion: "hard-selling, dispensing of misinformation, approaching women during labor, offering sterilization at a time of stress, on-the-job racism" ([7], p. 116).

Maria Diaz was disempowered in two ways. First, she was treated not simply without her consent but against her explicit will. In spite of her constant advocacy for herself, she did not participate even minimally in a very significant decision. Using Kapp's definition, she was disempowered.

But she was disempowered in another way as well. Sterilization is a life-changing operation with earth-shattering ramifications for women from "machismo" cultures. In "machismo" culture, a man's stature may be measured by the number of children he sires. He may divorce or abandon a woman who cannot bear children. Indeed, this is what happened to Lupe Acosta, whose common-law husband of eight years left her after she was sterilized against her will ([7], p. 107). She ended up on welfare, experiencing not only medical disempowerment, but social and economic disempowerment as well. A decision to refuse sterilization that may not seem "sensible" or "rational" in one culture may be very sensible in another. Empowerment in this context would require sensitivity to such cross-cultural issues, and recognition of the devastating consequences of what might seem a "sensible" decision in white North American culture.

Oliver Sacks may not have been told everything that he wanted to know, and may not have had the kind of communication that he desired. But at no time did he experience the kind of disempowerment that these poor, multiparous women with language and cultural barriers experienced. He was not treated *against* his will, nor was a foreign rationality imposed on him.

Medical Harm

Practices such as forced sterilization would be condemned as unjust by most observers. Harder to uncover are the injustices, the disempowerments, built into ordinary, routine medical practice. Here, there are two levels on which we must look for disempowerment and therefore for empowerment. In Sacks' case, the primary disempowerment came with the social *role* of patient – with losing control, being ignored, and not having social power.[4] But for many women, these social concomitants of the role of patient are only part of the picture. Medical practice historically has contributed not only to this second "misery" for women, but also to the first "misery," the phenomenon of physical disintegration itself.

Feminists and concerned women over the years have exposed a range of obstetrical and gynecological practices that actually *endanger* women's health. Unnecessary hysterectomies [15], use of the Dalkon shield [6], clitoridectomies [4, 23] – any number of practices with serious deleterious impact on women's health have been "routine" or common at one time in our history. For example, a number of studies have documented the movement from child-birth to the "delivery" of children in obstetrical units [11, 29]. Historical evidence suggests that midwifery was safer than obstetrics at the time when the (largely male) medical profession pushed out the (largely female) midwives. Many women have objected to the health risks presented for both mother and child by routine obstetrical practices. Ethel, who bore 16 children, puts it plainly:

I had all mine at home except the last six. . . . [I]t was easier to have them at home than to have them at the hospital. . . . They'd take me to the hospital and they'd strap me down. I'd like to never have the baby! When I was home, you know, I'd walk till the pains got so bad that I had to lay down then I'd lay down and have the baby. Without any anesthetic and never no stitches or nothing, because they waited till time. Now they cut you, you know, and they don't give you time to have it. . . . That's what ruins women's health ([3], p. 229).

That such practices actually are dangerous has been argued by several commentators [21, 25].

In this context, empowerment for women includes having control over one's own body and important medical decisions. The dimension of decision-making that Kapp lifts up remains important. But empowerment also includes better health care practices that do not endanger women or children.

In a study of court cases, Miles and August found that when the patient was a man, the court constructed his preferences for treatment in 75% of cases and allowed those preferences to be determining. But when the patient was a woman, the court constructed her preferences in only 14% of cases, and her preferences were not determinative of treatment decisions. Miles and August conclude that "women are disadvantaged in having their moral agency taken less seriously than that of men" ([20], p. 92). If Sacks had difficulty being heard and treated as a moral agent, imagine what he might have faced had he been a woman instead of a man. Thus, on the level of the second "misery," which Sacks calls the moral level, women are not treated equally with men. Empowerment in such a case means not having a double standard: both male and female patients should be treated with attention to their own expressed preferences as well as to their familial connections.

But it is not only the second level of "misery" on which women are not treated equally. Ayanian and Epstein found that "women who are hospitalized for coronary heart disease undergo fewer major diagnostic and therapeutic procedures than men" ([2], p. 221). Steingart et al. argue that it is "disturbing" to find that women report more cardiac disability before infarction than do men, but are less likely to receive treatments known (in men) to lessen symptoms and improve functional capacity ([26], p. 230). Just as Miles and August found that women's statements of not wanting to be kept alive were dismissed as "emotional" rather than "rational" desires, so it is possible that women's complaints of chest pain may not be taken as seriously as men's. Empowerment for women in the clinical setting clearly begins with having our voices honored and appropriate interventions utilized.

Empathy and the Non-treatment of Women's Issues

Another subtle form of disempowerment is the non-treatment of or non-focus on women's issues. Coronary artery disease is the leading cause of death in women ([26], p. 226). Yet our common image of "heart attack" is an image of a middle-aged professional man, not an image of a woman. The studies just cited make clear that there is much we do not know about how to treat heart disease in women. We have focused on men's needs, but not on women's.

Similarly, more women die *each year* of breast cancer than men died of AIDS in the first *ten years* of the epidemic [30]. It is estimated that one out of every three women will get cancer during her lifetime, and the breast cancer incidence rate has increased 32% in the last decade [27]. Yet, there is neither the commitment of funds for research and development of new treatments and interventions nor the commitment of public energies and attention to breast cancer that we currently experience for AIDS. Empowerment means *attending* to women's issues, and making them a priority for clinical research and treatment.

The reasons for the relative lack of attention to issues so central to women's lives are complex. Lack of women physicians and researchers may be an important contributing factor. A history of exclusion of women from top ranks of the medical profession leaves its legacy. Walsh concludes her historical study of the discrimination against women in the medical profession with these cautioning words: "There is an interrelationship between discrimination against women as medical students and physicians and against women as patients – resulting in the present lack of research on breast cancer, excessive rates of hysterectomies and surgery on women . . . and generally deficient health care for women" ([28], pp. 281–282).

Lack of women physicians not only influences choices about research and clinical emphasis; it also influences the possibilities for empathy, so important in Sacks' experience of healing. What male physician can truly empathize with the birth pains of a woman patient, or with what it means to a woman to lose a breast to surgery? If empathy is important for empowerment, then women will experience less empowerment than men when they are treated in a system that does not encourage women physicians. Similarly, white care-givers have difficulty empathizing with women and men of color; the well-to-do will not even imagine some problems experienced by those who are economically disadvantaged; and so on. Empowerment in the clinical context will require a change in the system that encourages different care-providers.

Problems of Access

While the focus of this essay is on empowerment *within* the clinical setting, some of the most important empowerment issues for those who are relatively powerless have to do with access *to* the clinical setting. Ectopic pregnancy is now the leading cause of maternal death among African-American women [19]. But in 1982 there were 44,000 women in New York state identified as at "high risk" who became pregnant, had no health insurance, and were not eligible for Medicaid ([19], p. 58). Without health insurance or the means to pay, these women do not get *into* the clinical setting. Their empowerment must begin outside that setting, with changes in political, social, and economic policies, Medicaid eligibility, and access to health care. Similarly, for older women, changes in Medicare to allow access to needed health care is critical.

While problems of access raise larger social and political issues, there are some things that can be done within the clinical setting to address these problems. For instance, would women who are eligible for Medicaid be able to find a physician who accepts Medicaid patients? "The worst medical problems I've had really," says Ethel, a poor mountain woman with 12 living children, "has been since I been on welfare. Trying to see the kinds of doctors that's needed for the children and myself, and they don't take the card – needing to see specialists, and the specialists don't take the card" ([3], p. 231). At the same time that we are experiencing the "feminization of poverty," with women increasingly among the poor who must depend on Medicaid, we are also experiencing a time in which ob-gyns, who specialize in women's diseases and reproductive processes, have the lowest rate of Medicaid participation of all primary care physicians ([19], p. 57). Women who cannot get into the health care system at all are doubly disempowered.

Empowerment in this context means the willingness of clinical care providers to "take the card" and deal with the government red tape, the bureaucratic form-filling, and the loss of income represented by accepting those patients. If more physicians "took the card" and had to deal with these inconveniences, perhaps we would see a faster move toward a more equitable system of access for the poor, many of whom are women.

Alternative Structures

Under these circumstances, it is no wonder that many women and other relatively powerless people have felt that empowerment cannot happen within the system. Empowerment means not only advocating for oneself and participating in significant decisions, but receiving care from a radically re-oriented system.

Some have moved to establishing alternative health care systems. One such organization was called "Jane" [1, 12]. Run by women on a non-hierarchical basis, "Jane" helped women get access to safe abortion during the time that abortions were largely illegal in the United States. "Jane" was part of a larger movement that involved teaching women about their bodies, their sexuality, and their own medical care [8, 9]. Alternative clinics were set up where women were trained to do their own vaginal examinations and to monitor their gynecological health. These organizations were empowering for women because they gave women knowledge, allowed women to help each other in non-hierarchical structure, and kept control of important bodily processes largely in the hands of women themselves. They strengthened relationships among women patients and providers, and tried to deal with issues that were central from the perspective and rationality of women of different cultures.

But alternative structures outside the system are not the only solution. Recognizing the rise of breast cancer and the crucial place of mammograms in diagnosis and early treatment, The Medical Center of Central Massachusetts set out to discover why women were so reluctant to come in for mammograms and whether something could be done about it. They asked women to talk about what keeps them from having mammograms. Among the factors that keep women away, they found these:
1. lack of child-care;
2. cold and unattractive hospital gowns;
3. lack of privacy;
4. inadequately trained technicians, with resulting pain and discomfort;
5. lengthy waiting time between testing and results.
The mammography unit was redesigned to address these problems: it now provides child-care, privacy, specially trained technicians, attractive and warm clothing, immediate test results, and so on. Such structural changes provide

empowerment not just for individual patients, but for the entire class of patients and ultimately for society as a whole.

Summary

The lesson to be learned from those who are relatively powerless is that we need changes in the system, not just changes of attitude in a few care-providers. More thoughtful listening, a willingness to hear the "whole" story and not just the "salient facts" will still be important. But it is the system that must be scrutinized for how it disempowers those who are already powerless, and how it could be made more empowering instead.

Empowerment in the clinical setting will require allowing patients to be their own advocates and to participate in significant decision-making. It will require not treating patients against their will, nor assuming that one culture's "rationality" makes sense for all cultures. It will require honoring those forms of rationality, such as art and religion, that offer a language "beyond" reason and science, a language that may be more appropriate to the patient's needs. It will require recognizing the network of community and relationships that affect patients' lives and decisions. But above all, it will require changing the system so that those who are relatively powerless have access to health care, for without that access, all talk of empowerment within the clinical setting is void.

Pacific School of Religion
Berkeley, California
U.S.A.

NOTES

[1] The definition of justice and dimensions of justice as empowerment are furthered explored in *Six Theories of Justice* [18] and *Justice in An Unjust World* [16].
[2] I have also heard him speak publicly and found his address strong.
[3] Precisely because these patients are already relatively powerless in the system, they are less likely to write books about their experiences than are the more powerful who become patients.
[4] Loss of control and autonomy is a typically male problem; feminist literature suggests that loss of relationship might be more problematic for women.

BIBLIOGRAPHY

1. Addelson, K. P.: 1986, 'Moral Revolution', in M. Pearsall (ed.), *Women and Values: Readings in Recent Feminist Philosophy*, Wadsworth, Belmont, CA., pp. 291–309.
2. Ayanian, J. Z., and Epstein, A. M.: 1991, 'Differences in the Use of Procedures between Women and Men Hospitalized for Coronary Heart Disease', *New England Journal of Medicine* 325(4): 221–230 (July 25).
3. Baker, D.: 1977, 'The Class Factor: Mountain Women Speak Out on Women's Health', in C. Dreifus (ed.), *Seizing Our Bodies*, Random House, NY, pp. 223–232.
4. Barker-Benfield, G. J.: 1977, 'Sexual Surgery in Late-Nineteenth Century America', in C. Dreifus (ed.), *Seizing Our Bodies*, Random House, NY, pp. 13–41.

5. Brueggemann, W.: 1989, *Finally Comes the Poet: Daring Speech for Proclamation*, Fortress Press, Minneapolis.
6. Dowie, M., and Johnston, T.: 1977, 'A Case of Corporate Malpractice and the Dalkon Shield', in C. Dreifus (ed.), *Seizing Our Bodies*, Random House, NY, pp. 86–104.
7. Dreifus, C.: 1977, 'Sterilizing the Poor', in C. Dreifus (ed.), *Seizing Our Bodies*, Random House, NY, pp. 105–120.
8. Frankfort, E.: 1977, 'Vaginal Politics', in C. Dreifus (ed.), *Seizing Our Bodies*, Random House, NY, pp. 263–270.
9. Fruchter, R. G., et al.: 1977, 'The Women's Health Movement: Where Are We Now?', in C. Dreifus (ed.), *Seizing Our Bodies*, Random House, NY, pp. 271–278.
10. Goldman, A. H.: 1980, *The Moral Foundations of Professional Ethics*, Rowman and Littlefield, Totowa, NJ.
11. Haire, D.: 1972, *The Cultural Warping of Childbirth*, International Childbirth Education Association, Seattle.
12. 'Jane': 1990, 'Just Call "Jane"', in M. G. Fried (ed.), *From Abortion to Reproductive Freedom: Transforming a Movement*, South End Press, Boston.
13. Jonas, H.: 1970, 'Philosophical Reflections on Human Experimentation', in P. Freund (ed.), *Experimentation with Human Subjects*, George Braziller, NY, pp. 1–31.
14. Kapp, M. B.: 1989, 'Medical Empowerment of the Elderly', *Hastings Center Report* 19(4), (July–August).
15. Larned, D.: 1977, 'The Epidemic in Unnecessary Hysterectomy', in C. Dreifus (ed.), *Seizing Our Bodies*, Random House, NY, pp. 195–208.
16. Lebacqz, K.: 1987: *Justice in An Unjust World: Foundations for a Christian Approach to Justice*, Augsburg Publishing House, Minneapolis, MN.
17. Lebacqz, K.: 1985, *Professional Ethics: Power and Paradox*, Abingdon Press, Nashville, TN.
18. Lebacqz, K.: 1986, *Six Theories of Justice: Perspectives from Philosophical and Theological Ethics*, Augsburg Publishing House, Minneapolis, MN.
19. McBarnette, L.: 1988, 'Women and Poverty: The Effects on Reproductive Status', in C. A. Perales and L. S. Young (eds.), *Too Little, Too Late: Dealing with the Health Needs of Women in Poverty*, Harrington Park Press, New York.
20. Miles, S. H., and August, A.: 1990, 'Courts, Gender and "The Right to Die"', *Law, Medicine, and Health Care* 18(1–2): 85–95 (Spring/Summer).
21. Rich, A.: 1986, *Of Woman Born: Motherhood as Experience and Institution*, W. W. Norton, NY.
22. Rosenbaum, E.: 1988, *A Taste of My Own Medicine*; now published as *The Doctor*, Ivy Books, NY.
23. Rothman, B. K.: 1979, 'Women, Health and Medicine', in J. Freeman (ed.), *Women: A Feminist Perspective*, 2nd ed., Mayfield Publishing Co., Palo Alto, CA.
24. Sacks, O.: 1984, *A Leg To Stand On*, Harper and Row, NY.
25. Sarah, R.: 1988, 'Power, Certainty, and the Fear of Death', in E. H. Baruch, A. F. D'Adamo, and J. Seager (eds.), *Embryos, Ethics, and Women's Rights: Exploring the New Reproductive Technologies*, Harrington Park Press, New York.
26. Steingart, R. M., et al.: 1991, 'Sex Differences in the Management of Coronary Artery Disease', *New England Journal of Medicine* 325(4) (July 25), 226–230.
27. Steingraber, S.: 1991, 'Lifestyle Don't Kill. Carcinogens in Air, Food, and Water Do', in M. Stocker (ed.), *Cancer as a Women's Issue: Scratching the Surface*, Third Side Press, Chicago.
28. Walsh, M. R.: 1977, *'Doctors Wanted: No Women Need Apply': Sexual Barriers in the Medical Profession, 1835–1975*, Yale University Press, New Haven.
29. Wertz, R. W., and Wertz, D. C.: 1977, *Lying-In: A History of Childbirth in America*, The Free Press, NY.
30. Winnow, J.: 1991, 'Lesbians Evolving Health Care: Our Lives Depend on It', in M. Stocker (ed.), *Cancer as a Women's Issue: Scratching the Surface*, Third Side Press, Chicago.

SECTION III

BEYOND PRINCIPLES

PAUL LAURITZEN

LISTENING TO THE DIFFERENT VOICES: TOWARD A MORE
POETIC BIOETHICS

> If detached, purely intellectual, abstract, deductive
> reasoning about universals was the winner in this
> transcendence-oriented tradition, it is easy, too, to
> see who the losers were. The losers were stories, and
> the storytelling imagination and emotions ([30], pp.
> 385–386).

In *By Blue Ontario's Shore*, Walt Whitman offers his vision of the public
role of the poet in a situation of conflict and diversity. The poet, he says,
may bring souls together in a way that coercion or a legal code ("paper and
seal") cannot. The poet can bind us together like "the limbs of the body or
the fibres of plants" because he has the "eye to pierce the deepest deeps.
. . ." [38], pp. 312–313). About the poet Whitman writes:

He is no arguer, he is judgment . . .
He judges not as the judge judges but as the sun falling round a helpless thing . . .
He sees eternity in men and women, he does not see men and women as dreams or dots. ([38],
     p. 313)

The thesis of this paper can be stated simply in relation to Whitman's observation. Bioethics has too frequently encouraged physicians to judge as the
judge judges and too seldom as the sun falling round a helpless thing. Because
so much work in bioethics has focused on a legalistic model of applying general
principles, for example, of autonomy, or justice, or beneficence, to troublesome cases, ethicists have not paid sufficient attention to dimensions of the
physician-patient encounter and of the process of moral decision making in
that encounter that cannot easily be subsumed under a judicial model.

One aspect of the clinical encounter that is obscured when bioethics is
excessively legalistic is the emotional dimension of the physician-patient
relationship. Since I believe it is crucial for bioethics to provide a place for
the emotional dimension of the clinical encounter and thus provide a place
for emotions in moral decision making, I also believe that we need to overcome
the legalism of bioethics. Thus, one way to summarize my central thesis is
to say that we need to make bioethics more poetic by allowing it to be more
passionate. We need, that is, to encourage a way of seeing patients, of being
with and for patients, that avoids the studied distance of the judge, that avoids
treating patients as dreams or dots, and instead treats them as concrete
individuals with particular hopes, dreams, fears, and frailties that inevitably
and properly call forth passionate responses from us.

The claim that contemporary ethical theory is too legalistic, too deduc-

151

G.P. McKenny and J.R. Sande (eds.), Theological Analyses of the Clinical Encounter, 151–169.
© 1994 *Kluwer Academic Publishers. Printed in the Netherlands.*

tive, too abstract is not new. Even the claim that bioethics is too focused on abstract rules and principles is not new. Indeed, at least since Carol Gilligan's book, *In A Different Voice*, was published in 1982, there has been a sustained attack on traditional moral philosophy for its attention to justice at the expense of care [1, 6, 13, 15, 29]. And the implications of this attack for bioethics have recently been examined as well [7, 17]. Nevertheless, advocates of a care perspective have not fully explored the importance of their critique of a justice orientation to recovering the significance of emotions in the moral life. One task for this paper, therefore, is to examine how the legalism of the justice perspective obscures the role of emotions in moral decision making generally and in bioethics in particular. Yet before we take up this task, we need to consider the general contours of recent criticism of contemporary ethical theory, including bioethics.

JUSTICE VS. CARE IN BIOETHICS

We can begin by briefly reviewing the critique of traditional moral philosophy that emerges out of Gilligan's work. As is well known, Carol Gilligan claimed to hear a distinctive moral voice in her studies of the moral development of women. According to Gilligan, women, unlike men, tend to see the world as a place of "care and protection" rather than of "dangerous confrontation" ([15], p. 38). As a result, women are less likely to think of morality as a matter of adjudicating conflict through the use of rules and principles intended to protect individuals from one another and to resolve disputes. Consequently, rights language, with its inherently adversarial character, has a less central place in women's moral vocabularies. Women talk instead of responsibility.

Indeed, Gilligan argued that we can discern two distinct approaches to morality by listening to the differences between men and women's moral voices. "The moral imperative that emerges repeatedly in interviews with women," Gilligan writes, "is an injunction to care, a responsibility to discern and alleviate the 'real and recognizable trouble' of this world." By contrast, she says, "for men, the moral imperative appears rather as an injunction to respect the rights of others and thus to protect from interference the rights to life and self-fulfillment" ([15], p. 100).

Gilligan's claim that we can identify two distinctive and typically gendered approaches to morality, a justice perspective and a care perspective, has been controversial [4, 6, 12, 28]. Nevertheless, it has fueled a substantial and fruitful critique of much of traditional moral philosophy. Following Gilligan, critics of traditional ethical theory have articulated their criticisms by highlighting a set of contrasts between what has preoccupied most moral philosophers and what they have neglected. They have celebrated rights, but neglected responsibilities; they have idolized detached calculation, but ignored affective attachments; they have attended to justice, but overlooked care. In short, traditional moral philosophy has simply been too detached, intellectual, abstract, and deductive [1, 13, 21, 29].

It is not hard to see how this criticism could be extended to encompass bioethics, for bioethics has arguably been preoccupied with articulating rules and principles that could be applied to moral quandaries in medicine [2, 8, 21, 26, 32]. Alisa Carse, for example, has explored the ways in which the contrasts between justice and care is directly relevant to bioethics [7]. She identifies four criticisms that care theorists make against a justice perspective, and shows how each can be extended to question standard approaches to bioethics. Since Carse's analysis draws attention to some of the problems with standard approaches to bioethics that I wish to explore below, let us turn briefly to consider her discussion.

The first criticism Carse identifies is that a justice perspective is obsessively focused on impartiality. Concerned first and foremost with fairness, justice theorists ask us to abstract from the particular circumstances of our separate lives in order to reason impartially and thus fairly. To be fair we must discount the significance of our personal attachments, life plans, and relationships, and adopt what Thomas Nagel has called the "view from nowhere." The problem with a preoccupation with impartiality from a care perspective is that the sort of abstraction to which it gives rise creates in turn a detachment and distance that ultimately leads to indifference. As Carse notes, quoting Gilligan, ". . . detachment, whether from self or from others is morally problematic, since it breeds moral blindness or indifference – a failure to discern or respond to need" ([7], p. 8).

The second criticism noted by Carse is that a justice perspective understands moral decision making primarily as a matter of applying general rules and principles to particular cases. Consistent with its concern for impartiality, justice theorists seek a sort of geometric model for moral decision making. One starts with rules and principles and reasons deductively to particular conclusions. According to Carse, the problem with this approach is that justice theorists frequently talk as if principles applied themselves. The justice perspective thus fails to account for the fact that principles must be applied and thus fails to provide an account both of how we recognize a situation as one to which particular principles are relevant, and of how precisely we apply principles to particular cases once we recognize the pertinence of particular principles to particular cases. Yet, to provide an account of how principles are properly applied is to move beyond an account of morality as essentially principle-governed, because the capacity for the appropriate attention and discernment of particulars necessary to apply principles is not itself derived from principles.

The third criticism Carse labels the "intellectualism" of the justice perspective. She has in mind the marked tendency of justice theorists to emphasize the use of reason in moral decision making and to define reason in opposition to emotion. Thus the intellectualism of the justice perspective consists in opposing the heart to the head and giving preference to the latter.

In doing so the justice perspective obscures the role emotions play both in the process of moral discernment and as a constitutive feature of appropriate moral response. From a care perspective, this concealment is the problem, for emotions often are central to moral discernment and *how* we respond to another in need may be just as important, indeed more important, than *what* we do in response to need. As Carse puts it, "through empathy or compassion, we may recognize another's pain or discomfort" ([7], pp. 13–14) and "expressing the right emotions at the right time in the right way is . . . an integral feature of moral agency" ([7], p. 14).

The fourth and final criticism is that a justice perspective is too individualistic. By focusing on a rights-based model of moral relations, a justice perspective highlights the norms of autonomy (to pursue individually defined conceptions of the good) and mutual noninterference. The problem with understanding moral relations after this fashion is that not all persons are equally capable of individually pursuing the good. Autonomy and mutual noninterference assume an equality that just does not exist. Once we acknowledge the inequalities between adults and children, the well and the ill, the rich and the poor, we see also that noninterference is a formula, not for flourishing, but for isolation and neglect.

As Carse points out, there are some fairly striking similarities between the traditional justice perspective and standard approaches to bioethics. Consequently, the four criticisms of the justice perspective that Carse identifies as rooted in an ethic of care can all be applied to work in bioethics without significant revision. For example, standard approaches in bioethics are largely principle-based and address the question of what morally ought to be done in some quandary situation. As Ronald Carson notes, the approach of positing general principles and then applying them to morally problematic medical situations "is well nigh ubiquitous in contemporary bioethics" ([8], p. 51). Nor can there be much question that bioethics has been decidedly individualistic and focused on autonomy. And Carse is not alone in warning about the dangers of fixating on autonomy. Daniel Callahan, for example, has worried about unqualified appeals to autonomy in bioethics. "We can already notice," he says, "some intimations of the unpleasant possibilities here: physicians who, far from treating us paternalistically, treat us impersonally and distantly, respecting our autonomy but nothing else" ([2], p. 41).

Thus, Carse's suggestion that both criticisms – that moral theory is too narrowly focused on rules and principles and that it is too individualistic – apply also to standard approaches in bioethics is certainly correct. So, too, is her claim that taking these criticisms seriously would have direct implications for how bioethics is taught and practiced. Nevertheless, important as these two criticisms are, I will focus instead on the other two criticisms Carse identifies and, in particular, on the relation between these two criticisms. For it is absolutely essential to see how and why a preoccupation with impartiality goes hand-in-glove with the eclipse of emotion in moral decision making and why the absence of attention to emotions in the clinical encounter is so troubling. Only when we explore these two criticisms and their relation to each other in detail, will we be in a position to see how bioethics might become less juridical and more poetic.

Carse's analysis provides us with a good start. She draws our attention to some of the ways in which an emphasis on impartiality is distorting, and she offers a suggestive account of why ignoring emotions is problematic. Nevertheless, we need a fuller and more finely-grained account of these difficulties. We need to examine: why impartiality has been thought to be so central; how a preoccupation with impartiality contributes to abstraction; why abstraction is so disastrous if emotions are to be assigned a legitimate role; and how emotions might be given a more central place in moral decision making if we reject the sort of abstraction that is characteristic of much contemporary work in bioethics.

We can begin with the question of why impartiality has come to be identified as the hallmark of moral decision making in modern moral philosophy. Let me suggest two distinct but related answers to this question. The first answer has to do with the prestige of science in the modern period. As a number of writers have suggested [16, 19, 25, 35], the emphasis on impartiality in contemporary moral philosophy is probably best traced to the fact that since the seventeenth century the moral philosopher has too frequently become a sort of scientist manqué. The reason that this has in turn contributed to a preoccupation with impartiality has to do with the fundamental shift in our basic understanding of the cosmos ushered in by the scientific revolution. The key feature of this shift was the rejection of a world view that understood the cosmos as a meaningful order and its replacement with the view of the cosmos as objective process. Once the world is objectified in this way, the proper methodological stance comes to be understood as one of neutrality and detachment. In order to understand the objective processes of nature, the investigator must himself be objective and thus rid himself of subjective values or emotions that might bias his investigation.

Whether this picture of the scientist as dispassionate investigator is plausible, the attraction of this picture to moral philosophy cannot be denied. Given the impressive accomplishments of the natural sciences in controlling and manipulating the world of nature, given the privileged epistemic status of scientific truth in the modern period, it was perhaps inevitable that philosophers would seek to emulate scientists in adopting a methodological stance of objectivity. Stanley Hauerwas has nicely captured the dynamic by which impartiality becomes central in contemporary moral philosophy when he writes:

At least partially under the inspiration of the scientific ideal of objectivity, contemporary ethical theory has tried to secure for moral judgments an objectivity that would free such judgments from the subjective beliefs, wants and stories of the agent who makes them. Just as science tries to insure objectivity by adhering to an explicitly disinterested method, so ethical theory tried to show that moral judgments, insofar as they can be considered true, must be the result of an impersonal rationality ([16], p. 16).

So one reason impartiality has been thought to be so central is that moral philosophy has sought a sort of scientific objectivity for morality as a way of grounding morality on apparently as sound a foundation as modern science. Morality would thus get a sort of prestige by association.

There is, however, a second reason why impartiality has been so attractive, and it has to do with what might be called the quest for invulnerability. Consider the quest for invulnerability in the quintessential advocate of impartiality, Immanuel Kant. Although Kant was certainly concerned to obtain from morality the sort of objectivity found in the sciences, he also sought to secure for moral agents protection against any unmerited and uncontrollable reversal of fortunes that might compromise moral integrity. In other words, he sought to make the morally good person invulnerable. No matter how unlucky he might be, no matter what fate might throw his way, the good person, for Kant, must not be vulnerable; his integrity must not depend on events beyond his control.

There are, of course, a number of ways in which the moral agent's integrity might be threatened. One particularly vexing way arises in relation to a possible conflict of commitments. If irreconcilable conflict is genuinely possible, a situation may arise in which an individual is confronted with a tragic choice. She may choose to do X or she may choose to do Y, and both X and Y exert a claim on her. In such a situation, a person may be compromised no matter how she chooses.

I have argued elsewhere [22, 23] that Kant pursued a two fold strategy in responding to this difficulty. On the one hand, Kant sought to limit the number of possible conflicts by restricting the range of legitimate claims that a moral agent must consider in reasoning morally. On the other hand, Kant sought to assimilate a conflict of duties to a conflict of beliefs and to argue that, just as it is irrational to suppose that two conflicting beliefs could both be true, so too would it be irrational to believe that two conflicting duties could both be binding. For our purposes here, we need only focus on the first part of the strategy, because it is Kant's effort to limit potential conflicts of duty that leads him, in part, to an emphasis on impartiality.

We can see the importance of impartiality to this project by considering the fact that Kant can effectively define out of existence a number of putative conflicts if he invokes a sharp dichotomy between moral and non moral value, and insists on the overriding significance of the former. That is precisely what Kant does, and in distinguishing moral from non moral value he insists on the centrality of impartiality to morality. Kant's insistence on impartiality thus helps to factor out from moral consideration the sorts of personal attachments to friends and family, and to one's own personal projects, that might otherwise generate substantial moral conflict. If, when deliberating morally, we must discount our particular attachments, say, to friends and family, then moral decision making is greatly simplified and the possibility of tragic conflict greatly reduced. Thus, just as impartiality was essential to the project of making moral judgment as rigorous as scientific judgment, so too is it essential in the quest for invulnerability.

In saying this, I do not mean to suggest that either the quest for invulnerability or that for scientific rigor can fully explain why impartiality has been a defining feature of modern ethical theory. Yet, even if these answers only partly account for the centrality of impartiality, they are a start, and they have one additional merit. They allow us to see why the concern for impartiality is so closely associated with the eclipse of emotions. If the model for moral deliberation has indeed been taken from the sciences, if the moral agent is to reason as a scientist reasons, it will come as no surprise that emotions will play no role in moral judgment. For the paradigm of the sciences is the dispassionate investigation. As Alison Jaggar puts it, the very logic of replicability in the sciences is "believed capable of eliminating or canceling out what are conceptualized as emotional as well as evaluative biases on the part of individual investigators" ([19], p. 162). Should we be surprised to discover, then, that a principle like universalizability is intended to accomplish the same thing in moral theory?

Moreover, those moral theories that have adopted a scientific ideal of objectivity have also typically wanted to supply a foundation for morality capable of supporting universal moral principles analogous to the universal principles of physics. Yet, as Mary Midgley has noted, the search for a foundation for morality is particularly inimical to emotions, especially when combined with a sharp dichotomy between reason and emotion. "The metaphor of foundation," she writes, "is disastrous; a building can only sit on one foundation, so it looks as if we have to make a drastic choice" ([27], p. 5). Not surprisingly, reason has always been the choice.

The quest for invulnerability is also inimical to emotion. As we saw, one sort of vulnerability for the moral agent arises out of a situation where duties or commitments conflict. Yet if emotions are treated as morally serious, if our emotional reactions are understood to be sources of genuine moral insight and to reflect genuine moral commitments, then the possibility for a conflict of duty or commitments is greatly expanded. Thus, for any theorist concerned to reduce or eliminate the possibility of tragic conflict and choice, one of the most significant advantages of the Kantian insistence on impartiality in moral deliberation will be that impartiality constrains personal attachments that give rise to potential conflict.

THE PROBLEM OF ABSTRACTION

I began this section of the paper by asking why impartiality has been thought to be central to ethical theory. In offering two admittedly partial answers to this question I have also provided a general sketch of why appeals to impartiality are so frequently combined with disdain for emotion. There are, however, more specific reasons why the appeal to impartiality poses obstacles to recovering the importance of emotions and, if we are to provide a place for emotions in bioethics, we must attend to these more specific barriers. The problems I have in mind cluster around the frequently alleged connection between impartiality and abstraction. Alisa Carse's statement of this connection is fairly typical: ". . . the rejection of impartiality as the mark of the moral," she writes, "is a rejection of the prevailing tendency in ethical theory to construe, as morally paradigmatic, forms of judgment that abstract away from concrete identity and relational context, and to view moral maturity and skill as residing essentially in the capacity for abstract judgment so construed" ([7], p. 10). So one problem with an insistence on impartiality is that it leads to abstraction. But what is the precise nature of the complaint against abstraction?

The best discussion of this question that I know is found in Owen Flanagan's recent book, *Varieties of Moral Personality*. In a chapter entitled, "Abstraction, Alienation, and Integrity," Flanagan takes up a set of related complaints that has been pressed against some versions of contemporary moral theory. The core complaint is that most versions of Kantianism and utilitarianism require an unrealistic degree of impartiality and thus require moral agents to abstract too heavily from their identities. This sort of abstraction it is argued can in turn lead to alienation from the agent's most central commitments and thus to a violation of integrity. In order to assess this complaint, Flanagan attempts to define it in some detail. His discussion is instructive.

According to Flanagan, we should begin by noting that abstraction is primarily a matter of cognitive processing and is a feature of all practical deliberation. Deliberation in any cognitive domain requires some degree of factoring out features of the self and the world that are irrelevant to the tasks at hand. So to say that abstraction is a central feature of some ethical theory is not in itself a serious objection to that theory. It becomes a serious objection only if the critic can show that what is being factored in or out is objectionable. In order to see whether critics of Kantianism and utilitarianism can sustain this latter claim, we must consider the ways in which the factoring process of abstraction might be defective.

According to Flanagan, there are two levels on which we can identify problems with abstraction and they correspond to two types of factoring. The first type of abstraction Flanagan calls "feature detection" or "classificatory abstraction." As the name suggests, this type of abstraction involves information detection and classification. It is an active process by which we isolate properties of an object or event and classify the object or event as a member of a type. The second type of abstraction Flanagan labels "task-guided abstraction." This form of abstraction goes beyond classificatory feature detection and "involves deployment of rationalized procedures deemed appropriate to the successful completion of the task at hand" ([11], p. 85). These procedures will assign various features of an object or event different weights and will underwrite a differential response to these objects and events on the basis of the assigned ranking.

With this as background, Flanagan argues that the objection that a moral theory requires an unacceptable level of abstraction can take either of two forms. It can be an objection to the effect that certain classifications warranted by the theory depend on limiting the number of salient features picked out in the process of classification. Or more commonly, as with many responses to Kantianism and utilitarianism, it can be an objection that the rules guiding abstraction give some saliencies too much weight and others too little. As Flanagan puts it, the process demanded by the theory either "fails to notice the relevant saliencies in the first place, or it notices them and than factors them out or weights them incorrectly" ([11], p. 86).

If we now return to the question of how, specifically, the preoccupation with impartiality and abstraction has contributed to the eclipse of emotions, we can see that the answer may, like objections to abstraction in general, take two forms. The most obvious form is an objection to the task-guided abstraction of a justice perspective. On this view, the problem with the justice perspective is precisely that it provides no place for emotion in its rationalized procedures for moral deliberation. On the contrary, in deciding morally, the agent is guided by a rule to the effect that he or she should factor out or, at best, discount the value of emotional response.

It is this sort of objection, I believe, that Carse had in mind when she wrote, in the passage noted above, that the rejection of impartiality on a care perspective is a rejection of a view of "moral maturity and skill as residing essentially in the capacity for abstract judgment." The alleged skill is that of discounting emotional response and other factors of quandary situations that compete with selective attention to the generic features of objects or events that might be relevant to identifying the application of rights. And there can be no question that this sort of skill, and its corresponding task-guided abstraction, is highly valued in bioethics. As Sidney Callahan points out, you do not have to search the bioethics literature long to find writers who recommend a decision making procedure that factors out emotions. Emotions are dismissed as merely visceral and bioethicists are urged "to become impartial reasoners 'whose only interests are in the consistency and force of rational argument'" ([4], p. 9).

We have already examined some of the reasons why moral theorists might be attracted to an account of moral deliberation that factors out or deeply discounts emotions in the process of moral decision making. So we should not be surprised that the task-guided abstraction required in Kantian or utilitarian theories leaves little room for emotions. It is important to see, however, that it is not just the task-guided abstraction that is the problem, but also that of feature detection. Once again, Flanagan's discussion is quite helpful.

As Flanagan points out, when critics claim that Kantian or utilitarian theories demand an unreasonable degree of abstraction of the moral agent in the process of moral decision making, they almost always have in mind a complaint about the task-guided abstraction of these theories. In other words, the complaint is usually that Kantian or utilitarian theories require agents to factor out too many relevant features of moral situations, features like personal attachments, life-plans, or, as we have just seen, emotions. Unfortunately, such complaints too frequently ignore the connection between task-guided abstraction and feature detection. Thus, says Flanagan, critics often appear to assume that an impartial deliberator "is as sensitive as any other person to the multifarious particularities of various situations. It is just that she factors most of this information out for purposes of moral deliberation" ([11], p. 86).

The problem with this assumption, however, is that there is good reason to believe that the ability to discern morally salient features of situations that require moral deliberation is itself shaped by how the agent understands the process of moral deliberation. In one sense, this point is simply an elaboration of Carse's criticism that principles do not apply themselves. Yet once we acknowledge the obvious fact that an adequate moral theory will need to offer not just a decision making procedure but some account of the process of discernment by which agents identify situations to which the procedure is applied, we see also that the one affects the other. As Flanagan puts it: "It is extremely unlikely that one could become proficient at recognizing moral challenges as defined by the lights of such a [Kantian or utilitarian] conception without its changing the way the world is *seen*" ([11], p. 87). It follows that any complaint against the task-guided abstraction of a moral theory is likely to be incomplete unless it is also accompanied by an account of the problem such abstraction raises for the process of moral discernment.

We can see this clearly in relation to the difficulties raised by the process of abstraction for accounting for the role of emotions in moral decision making. The problem is not just that the canonical decision making procedure for justice theories requires factoring out our emotional reactions to the situations about which we are deliberating. The problem is that adopting this sort of required distance from our emotional responses when we deliberate morally, will almost certainly also lead us to ignore the emotional cues that are such an important part of moral discernment in the first place. The problem, therefore, is not just that the process of factoring away emotional response when deliberating morally leads to a sort of woodenness in applying rules and principles. The more significant problem is that eliminating emotional response itself contributes to a sort of moral blindness. Martha Nussbaum has put this point forcefully. "A person of practical insight," she writes:

will cultivate emotional openness and responsiveness. . . . Frequently, it will be her passional response, rather than detached thinking, that will guide her to the appropriate recognitions. "Here is a case where a friend needs my help": this will often be "seen" first by the feelings that are constituent parts of friendship, rather than by pure intellect ([30], p. 79).

At this point, we can see that a preoccupation with impartiality is likely to impoverish moral decision making in two related ways. First, in so far as focusing on impartiality leads to a decision making procedure that requires factoring out emotional response in moral deliberation, it cuts us off from an important source of moral insight. Second, continual damping of emotional response is likely to result in a sort of woodenness in relating to others that itself frustrates the process of moral discernment. So the call for strict impartiality is related to the eclipse of the role of emotions in moral decision making, and extirpating emotions from moral deliberation makes the process itself a vastly poorer thing.

ABSTRACTION IN BIOETHICS – THE LOSS OF EMOTION

If focusing on impartiality has these consequences for moral theory generally, it should not be surprising to discover the same dynamic at work in bioethics when objectivity, impartiality, and abstraction are the watchwords for this discipline. To see how this is the case I turn now to consider two recent articles about the process of moral decision making in bioethics. Although neither article is primarily concerned with the importance of emotional response in decision making in bioethics, each illustrates how the abstraction characteristic of much work in bioethics frustrates rather than fulfills moral deliberation.

The first article is "Rich Cases – The Ethics of Thick Description," by Dena Davis. The essay is concerned, in part, with the difficulties facing bioethicists of reporting and commenting upon actual cases in a way that protects the privacy of patients. It is a reflective essay occasioned by the author's own effort to describe, for an article she was writing, four cases she had observed as a bioethics fellow in a way that preserved the privacy of the patients involved. She discusses the paradox she confronted that "the very details that needed to be falsified [in order to protect confidentiality] were just those that gave the cases their integrity and usefulness" ([10], p. 12). This paradox leads her to consider the use of cases in bioethics and to identify three functions cases serve. Cases may:

a) be a useful "tool for teaching theory" [12];
b) provide a pool of shared experience" [13]; or
c) serve as an experience in relation to which "ethicists can make points and draw conclusions" [13] and invite others to do likewise.

Davis notes that, with the exception of cases that serve the second function, most cases used in bioethics are "thinly" described. Unlike those cases where individuals or families have chosen to tell their stories fully and publicly, or where the details of a case are a matter of public record because the cases ended up in the courts, the typical case presented in ethics rounds or in commentaries or in textbooks in bioethics is devoid of the concrete details that make the former cases – like those of Dax Cowart or Nancy Cruzan – so helpful and so compelling. To be sure, cases described "thinly" protect

confidentiality in a way that is difficult if they are "thickly" described, but
Davis argues cogently that we must not abandon the effort to reason morally
in relation to the finely-textured particulars of real cases, because to do so is
to truncate and thereby to mutilate the process of moral reflection.

It is at this point that her article is most useful to us, for Davis suggests
that the penchant for thin description is typically associated with a rights-based
or justice approach to bioethics that is concerned fundamentally with universalizability, and she beautifully illustrates what is lost when such a
perspective is not complemented by a care perspective that refuses to factor
out the details of a case that so frequently call forth a personal response.

She recounts her efforts to describe a case in which a woman was left
comatose after an aneurysm burst during childbirth. How could she alter or
eliminate features of the case in order to protect privacy, but not fundamentally change how she or others would respond? She could change the fact
that the man who was left to fight so that his spouse be allowed to die had
been her husband for many years. She could change the fact that this couple
had struggled for many years to achieve this pregnancy that had ultimately
been disastrous for them. She could change the facts to give the couple other
children at home. All these changes, and others that could be imagined, would
help to disguise the actual identities of the parties in the case, but so too
would they change how the case strikes us.

If changing the details in these apparently small ways can fundamentally
alter how we approach cases morally, factoring out the details can disfigure
the case beyond recognition and utterly stifle the process of moral reflection
about such cases. Davis illustrates this point by contrasting how a justice
perspective might approach this case and how a care perspective might.
"Described thinly," Davis writes:

the case was "about" a doctor who was known for his refusal to give up, and who was blocking
the husband's attempt to respect his wife's autonomy by honoring her wishes. Described in terms
of an ongoing moral narrative (i.e., thickly), it was about the heroism of a man with a newborn
baby at home (the culmination of years of unsuccessful attempts at pregnancy) and a comatose
wife in the hospital, valiantly taking on the medical establishment in an attempt to be her
spouse one last time in making sure her wishes were honored. A rights-based, universal ethics
focuses on respect for the woman's autonomy as the primary issue. An ethics centered in
narrative and relationship focuses on the husband as the appropriate individual to convey her
wishes (due to the length of their marriage, quality of the relationship, and so on) ([10],
p. 14).

Davis rightly insists here that emphasizing the detailed narrative need not
mean giving up concern for autonomy and rights, that justice and care are complementary, not mutually exclusive, and we should not lose sight of this fact.
At the same time, her comments help illustrate why the abstraction that is characteristic of work in bioethics that focuses only on thinly described cases
requires us to factor out too much. In other words, we can see in Davis' discussion of this case the dramatic illustration of what is lost when a commitment
to impartiality leads us to dismiss emotional response as irrelevant to moral

deliberation. In this case, to prescind from the fact that this pregnancy was the culmination of years of struggle, that the couple, say, understood the child as the deepest fulfillment of an already happy marriage, to ignore the pain and ambivalence, possibly the horror, for the husband of having a child beginning life at home and a spouse ending life in a hospital, is to lose everything we need to respond to this case in a fully human way.

If Davis' essay helps us to see that approaching cases in bioethics solely from a justice perspective is unacceptable because it results in factoring out precisely those details that make a fully human response possible, the second essay, to which I now turn, illustrates how a justice perspective may lead us to overlook or distort such details in the first place. The essay is a case story and commentary by Warren Reich published in *Second Opinion*.

Reich describes the situation of a woman named Denny whom he interviewed in connection with an ethics committee consultation. Denny's seventy-two-year-old father, Jimmy D., was hospitalized due to a failing lung and was now dependent on a mechanical ventilator. The ethics committee had been asked to make a recommendation about whether Jimmy should be discharged to home care with his daughter; kept in the intensive care unit; or transferred to a long-term-care facility. After discussing the prospects of home care with Denny, the attending physician, Dr. Louis Robert, had recommended against discharging Jimmy to his daughter's care. Reich reproduces Dr. Robert's report to the committee.

What convinced me that we shouldn't trust her with her father's care was when she said to me: "I want to take care of my father; it's my responsibility, and I think I'm prepared to do it, but how can I care for him if no one helps me?"

I had already explained to her several times how she could get the help she needed; so I finally concluded that she was just using lack of help as an excuse for her unwillingness or inability to take on such a big responsibility. . . . ([32], pp. 41–42).

In explaining why he reached the opposite conclusion, Reich describes in detail the person he interviewed. She is, he says, a person whose whole life has been oriented toward caring for others. In a two-hour-long interview Denny tells Reich of her decision to become a parent out of a desire to care for a baby and to have someone she could teach to love and care for others. She tells him of the support of her husband and her sister that enabled her to care for her son who died at age twenty of cancer. She tells him of her sense of failure over the breakup of her marriage after the death of her son, a failure she perceives as the result of her inability to care fully for her husband.

Reich uses Denny's case as a point of departure for a very rich discussion of the nature and meaning of care, of the role of compassion in a caring situation, and of the resources required to enable care. He also uses Denny's case to illustrate the contrast between a justice and a care perspective. According to Reich, Dr. Robert understood this case in fundamentally different terms than did Denny, and he used a different moral vocabulary in characterizing the case. In Reich's words, "Dr. Robert viewed the care of Denny's father through the lenses of justice and legal adversarialism, Denny saw it as a situation of responsiveness in a relationship of mutuality with her father" ([32], p. 51). Where Dr. Robert understood the moral issue raised by the case as how best to discharge a duty to care, understood primarily as the delivery of a service, Denny approached the case as a situation demanding a virtuous personal response from her; not the delivery of a service, but rather a personal presence.

Reich shows how this difference in perspective ramifies in any number of ways throughout the presentation of this case to the committee. There is, however, one consequence to which Reich draws our attention that is particularly relevant to our purposes, for it illustrates how the abstraction of a justice perspective may lead to a failure of discernment. According to Reich at one crucial juncture, Dr. Robert simply fails to understand Denny, precisely because he sees through "the lenses of justice and legal adversarialism." According to Reich, "Dr. Robert unknowingly misrepresented Denny's words to the committee."

Perhaps because his was the world of the justice perspective, he reported her words this way: "How can I take care of my father if no one *helps* me?" He interpreted and reshaped her words as a fairness statement – a sort of quid pro quo – as though she was refusing to give aid to her father unless she got material and technical assistance. He simply didn't hear her moral voice, possibly because it was so different from his own ([32], p. 52).

The contrast between Professor Reich's response to Denny and Dr. Robert's response, illustrates what may be lost when a concern for justice leads us to approach cases abstractly and legalistically. Concerned primarily with discharging his duty to provide sufficient technical support to allow Jimmy's medical needs to be met, and concerned with the legal consequences of how this duty was discharged, Dr. Robert fails to respond to Denny's plea for emotional support. Dr. Robert recognizes that Denny's father wants to be cared for by his daughter and that his daughter wants, on some level, to take care of him. Consequently, Dr. Robert feels that it is "unfortunate" that he cannot recommend this course of action. But he factors out this response by describing it abstractly in terms of "accepted principles of medical ethics" involving ordinary and extraordinary care. Approaching the case abstractly he then misinterprets Denny's plea for help as a sort of *quid pro quo*.

By contrast, Professor Reich refuses to approach this case abstractly and legalistically. He tells us that he abandoned "the mandated interview process in favor of a simple conversation." He thus meets Denny first and foremost as a person, and not as a subject of a formal review. He refuses to translate her comments into principles of bioethics. He attends to her tone of voice, to the emotional rhythm of their conversation. He notes that at one point in the conversation she is "clearly tired," at another that she continues with "heavy resignation." He is clearly sympathetic and refuses to ignore or discount his own emotional responses to Denny as he, like Dr. Robert, tries to reach a conclusion about the case. Unlike Dr. Robert, however, Prof. Reich can see that Denny's call for help was not a self-deceptive excuse, but rather a legitimate expression of concern that she have some emotional support through the difficult time that awaits her. Unfortunately, because Dr. Robert was focused on professional duty and legal responsibility narrowly construed, he failed to see Denny's plea for what it was.

TOWARD A MORE POETIC BIOETHICS

Not only do these two cases thus help us to see the difficulties that arise for bioethics when emotional response is discounted in moral decision making, they also suggest how bioethics could provide a more central place for emotions in moral decision making. For although these two essays set out to make quite different points, they are united by a common theme, namely, the importance of storytelling to the process of moral deliberation. What both Reich and Davis highlight is how an exclusive concern with principles or rights or justice flattens out the rich texture of the stories that shape the lives of both patients and physicians. What Reich's essay in particular reveals is how such a process of flattening out the narrative structure of these experiences also flattens out the emotional responses of the parties involved in a way that is highly deleterious to moral decision making.

In another essay [31], Reich calls for the development of "experiential ethics." "An experiential ethic," he says, "begins with a perception and interpretation of values related to moral experience – person's relationships, roles, attitudes, behaviors – that are conveyed through life experiences, narratives, images, models known from behavioral sciences, etc" ([31], p. 283). Such an ethic, I believe, would in fact make bioethics more poetic; it would provide a place for the emotions that are so crucially a part of the moral response to decline and death, our own and others.

To reshape bioethics to be more attentive to the narratives of patients and their families will require not only thick description of the cases by which bioethics is typically taught, it will require educating and training physicians (and others) to attend to the "nonmedical" stories that patients tell [37]. It will require helping physicians not to flee to the disengaged abstraction of the medical history or to the applications of principles of bioethics, thereby pushing "the person of the patient into the background" ([9], p. 173). We must move, in other words, from a medical model of the interaction between physician and patient in the clinical encounter "which distances and protects" the doctor "from direct interaction with painful situations and feelings, to one that supports his ability to remain both intellectually and emotionally present" ([37], p. 56).

To urge all those concerned with bioethics to become emotionally present to patients and their stories is not to advocate a model of physician as friend rather than that of physician as stranger. On either model a physician could retreat from the messy reality of the patient's story to the safety of approaching cases abstractly and distantly. Instead, to call for greater emotional aware-ness and responsiveness in the clinical encounter is to call for expanding our conception of bioethics to encompass the development of skills of discernment that are certainly necessary to the application of principles, but go beyond the mere application of principles. It is to call for the cultivation of emotional capacities to respond passionately to patients and their families by remaining close to and moved by the concrete details of their stories ([34], p. 168).

If bioethics were to move in this direction, we could expect a number of changes in the discipline. One change is that bioethics might well become more poetic, not simply in the metaphorical sense implied by the title of this essay, but more literally through the study of narrative forms that demand attention to detail and call forth passionate personal responses. Indeed, a number of writers have already pointed to the power of literature to sharpen and intensify our appreciation both of the experiences of others with whom we may typically have little contact and of the experiences of those near to us, including ourselves, whose very familiarity may cloud perception and deaden response [9, 18]. As Martha Nussbaum has noted, literature may "offer a distinctive patterning of desire and thought" because a novel, say, may "ask readers to care about particulars, and to feel for those particulars a distinctive combination of sympathy and excitement" ([30], p. 236). We might add, "or a distinctive combination of disgust and revulsion," but her general point is surely correct, and would be acknowledged and incorporated into the study of medicine given an expanded understanding of bioethics.

We have, of course, already seen one way in which such insight might be incorporated in bioethics, for Dena Davis' article highlights how thick description is possible in bioethics. And thick description is precisely what a novel offers as it cultivates emotional capacities in the readers who come to care about the characters and their stories. So we could expect a greater emphasis on thick description of actual cases drawn from bioethics as well as a recognition of the potential usefulness of thick description of fictional cases in the process of moral deliberation. In my view, such a change would indeed be significant.

There is, however, a more far-reaching consequence of recovering the importance of emotional response to bioethics, and I want in closing to draw attention to it, even while acknowledging that a more thorough discussion is necessary. We can see this particular outcome of recovering the proper role of emotions in moral decision making by considering once again the connections I have drawn throughout this paper between the process of abstraction and the eclipse of emotion. I have suggested above that to deny our emotional attachments or to dismiss them when reasoning morally is to cut ourselves off from an important source of moral insight, and that this is the upshot of a preoccupation with impartiality and abstraction. Yet, if dismissing emotions puts us at odds with some of our deepest value commitments, it is not because emotions are blind "pushes and pulls," but because emotions are experiences that manifest and reflect our sense of ourselves and our situations. As such, emotions are deeply embedded in the stories that define who we are and what we care about. This in turn suggests that our emotions are much more deeply rooted in religious belief and practice than is usually supposed, at least by those writers who seek to abstract from such belief and practice in the quest for objectivity.

Thus, to ignore or discount emotional response is not merely to impoverish practical deliberation, it is likely also to obscure what Clifford Geertz ([14], p. 126) has identified as the synthesis of "ethos" and "world view" in a group's sacred symbols. As Geertz has noted, there is a dialectical relationship between religious world views and emotions. On the one hand, religious belief and practice gives order and definition to our emotions, on the other hand, our affective life, thus structured by our understanding of the way the world ultimately is, reinforces our world view by making it emotionally convincing.

The significance of this point to debates in bioethics is that it requires us fundamentally to rethink the role of religious belief to these debates. For example, more than one observer has commented on the tendency of religious writers to present their views in fundamentally secular terms. As Allen Verhey trenchantly remarks, some Christian bioethicists are "more easily identified as followers of Mill or Kant than as followers of Jesus" ([36], p. 22). Such attempts to translate explicitly religious convictions into terms that are readily accessible to all is perhaps understandable, but we should not loose sight of the fact that this sort of translation is itself a kind of abstraction that will be as corrosive to emotional response as other, more obvious, forms of abstraction.

Indeed, the assumption often driving such efforts of translation is that the conclusions are all that ultimately matter, that if we agree about what ought to be done, how we reached agreement is of little consequence. Although there is no gainsaying the importance of agreement on public policy issues in a pluralistic democracy, we must not lose sight of the fact that purely cognitive agreement may in fact yoke together fairly disparate moral responses. We have already noted, for example, that how we respond may be as important as what we do in response. So agreement about what we should do may mask significant differences. Yet, even this way of putting the point is a bit misleading, because it suggests that emotional response is something extra that ideally might be added to agreement in principle but is not generally necessary. The problem with understanding agreement in this way is that it seeks to treat pieces of a web as if they were unconnected to a larger whole. Nancy Sherman has made this point eloquently in relation to the moral response of compassion. "The evaluative content of emotions," she writes:

may not be purely cognitive or intellectual. To respond compassionately to a loved one who is suffering may not simply be a matter of (intellectually) seeing, and feeling compassion as a result, nor conversely of seeing *because* one feels compassion, but of seeing with an intensity and resolution that is itself characterized by compassion. One would not have seen in *that* way unless one had certain feelings. The mode of seeing is distinct ([34], pp. 170–171).

The problem with translating religiously rooted approaches to bioethics into secular terms is thus not simply that, once translated, the religious contribution would be indistinguishable from that of non-religious contributions, "just so much more philosophy," as H. Tristram Engelhardt, Jr., puts it. The problem is that translating the detailed narratives of religious traditions will rob those narratives of their power to give order and definition to our emotions and thus sever the connection between seeing and feeling. Once this connection is severed, however, the mode of seeing may be quite different. So the effort to translate religious responses in order to make them accessible may in fact have the opposite effect. It may, for example, render the passionate responses of religious believers particularly opaque, because these responses will thereby be sundered from the story that gave them shape and meaning.

The worry that arises if we eschew translation, of course, is that we condemn religious discourse to a completely marginal status in public policy debates, because it will be incomprehensible to so many in the debate. This view seems to me unduly pessimistic. Yet, even if religious traditions remain somewhat opaque, even if the religious voices in the public conversation of bioethics are not fully comprehensible to everyone in the debate, we must not distort these voices by rendering them in the flat, affect-less, technical language of traditional bioethics. At least we must not do this if we want bioethics to become more poetic.

We saw at the start that Whitman describes the judgment of the poet as like the sun falling round a helpless thing. Elsewhere he reflects on the process involved in such judgment. Sometimes, he says, the poet must just stop and listen.

Now I will do nothing but listen,
To accrue what I hear into this song, to let sounds contribute toward it.
I hear bravuras of birds, bustle of growing wheat, gossip of flames, clack of sticks cooking
    my meals,
I hear the sound I love, the sound of the human voice . . . ([38], p. 85).

For bioethics to become more poetic, it must learn to love the sound of the human voice, and it must try not to distort that voice either by robbing it of its passion or by insisting that it speak in a familiar tongue.

John Carroll University
Cleveland, Ohio
U.S.A.

BIBLIOGRAPHY

1. Baier, A.: 1987, 'The Need for More than Justice', in M. Hanen, and K. Nielsen (eds.), *Science, Morality and Feminist Theory*, University of Calgary Press, Calgary.
2. Callahan, D.: 1984, 'Autonomy: A Moral Good, Not a Moral Obsession', *Hastings Center Report* 14(5), 40–42.
3. Callahan, D., et al. (eds.): 1985, Applying the Humanities, Plenum Press, New York.
4. Callahan, S.: 1988, 'The Role of Emotion in Ethical Decisionmaking', *Hastings Center Report* 18(3), 9–14.
5. Callahan, S.: 1991, 'Does Gender Make a Difference in Moral Decision Making?', *Second Opinion* 17(2), 67–77.

6. Card, C. (ed.): 1991, *Feminist Ethics*, University Press of Kansas, Lawrence, Kansas.
7. Carse, A.: 1991, 'The "Voice of Care": Implicatons for Bioethical Education', *The Journal of Medicine and Philosophy* 16, 5–28.
8. Carson, R.: 1990, 'Interpretive Bioethics: The Way of Discernment', *Theoretical Medicine* 11, 51–59.
9. Cassell, E.: 1985, 'The Place of the Humanities in Medicine', in D. Callahan, et al. (eds.), *Applying the Humanities*, Penum Press, New York.
10. Davis, D.: 1991, 'Rich Cases: The Ethics of Thick Description', *Hastings Center Report* 21(4), 12–17.
11. Flanagan, O.: 1991, *Varieties of Moral Personality*, Harvard University Press, Cambridge, MA.
12. Flanagan, O., and Jackson, K.: 1987, 'Justice, Care and Gender: The Kohlberg-Gilligan Debate Revisited', 97, 622–637.
13. Friedman, M.: 1987, 'Beyond Caring: The De-Moralization of Gender', in M. Hanen, and K. Nielsen (eds.), *Science, Morality, and Feminist Theory*, University of Calgary Press, Calgary.
14. Geertz, C.: 1973, *The Interpretation of Cultures*, Basic Books, New York.
15. Gilligan, C.: 1982, *In a Different Voice*, Harvard University Press, Cambridge, MA.
16. Hauerwas, S.: 1977, *Truthfulness and Tragedy*, University of Notre Dame Press, Notre Dame, IN.
17. Holmes, H.B. (ed.): 1989, 'Feminist Ethics and Medicine', Focus Issue of *Hypatia* 4(2).
18. Hunter, K.H.: 1985, 'Literature and Medicine: Standards for Applied Literature', in D. Callahan, et al. (eds.), *Applying the Humanities*, Penum Press, New York.
19. Jaggar, A.: 1989, 'Love and Knowledge: Emotion in Feminist Epistemology', *Inquiry* 32, 151–176.
20. Kittay, E.F. and Meyers, D.T. (eds.): 1987, *Women and Moral Theory*, Rowman and Littlefield, Savage, MD.
21. Ladd, J.: 1979, 'Legalism and Medical Ethics', *The Journal of Medicine and Philosophy* 4, 70–80.
22. Lauritzen, P.: 1989, 'A Feminist Ethic and the New Romanticism – Mothering as a Model of Moral Relations', *Hypatia* 4, 29–44.
23. Lauritzen, P.: 1991, 'Error of an Ill-reasoning Reason: The Disparagement of Emotions in the Moral Life', *The Journal of Value Inquiry* 25, 5–21.
24. Lauritzen, P. 1992, *Religious Belief and Emotional Transformation*, Bucknell University Press, Lewisburg, PA.
25. MacIntyre, A.: 1981, *After Virtue*, University of Notre Dame Press, Notre Dame, IN.
26. May, W.F.: 1991, *The Patient's Ordeal*, Indiana University Press, Bloomington, IN.
27. Midgley, M.: 1981, *Heart and Mind*, St Martin's Press, New York.
28. Nails, D., M.A., O'Loughlin, M.A., and Walker, J.C.: 1983, 'Women and Morality', *Social Research* 50 (Focus Issue).
29. Noddings, N.: 1984, *Caring: A Feminine Approach to Ethics and Moral Education*, University of California Press, Berkeley.
30. Nussbaum, M.: 1990, *Love's Knowledge*, Oxford University Press, New York.
31. Reich, W.T.: 1987, 'Caring for Life in the First of It: Moral Paradigms, for Preinatal and Neonatal Ethics', *Seminars in Perinatology* 11, 279–287.
32. Reich, W.T.: 1991, 'The Case: Denny's Story,' and 'Commentary: Caring as Extraordinary Means', *Second Opinion* 17(1), 41–56.
33. Shelp, E. (ed): 1983, *The Clinical Encounter*, D. Reidel, Dordrecht.
34. Sherman, N.: 1989, *The Fabric of Character*, Oxford University Press, New York.
35. Taylor, C.: 1985, *Human Agency of Language*, Cambridge University Press, Cambridge.
36. Verhey, A.: 1990, 'Talking of God – But With Whom?' *Hastings Center Report*, Supp, 20(4), S21–24.
37. Warshaw, C., and Poirier, S.: 1991, 'Case and Commentary: Hidden Stories of Women', *Second Opinion* 17(2), 48–61.
38. Whitman, W.: 1968, *The Works of Walt Whitman*, vol. 1, Funk and Wagnalls, New York.

SANDRA W. CHURCHILL AND LARRY R. CHURCHILL

REASON, NARRATIVE AND RHETORIC:
A THEORETICAL COLLAGE FOR THE CLINICAL ENCOUNTER

INTRODUCTION

Ethical Theory as Collage

Human ethical activity is complex and multi-faceted. It includes deciding and acting, reflection and contemplation, sustaining habits over time, nurturing character, and many other kinds of activity. It is part logic, part acts of will, part turning of emotional sensibilities, part feats of imagination. Because ethics embraces such diverse kinds of activity, ethical theories need to be diverse and multi-faceted as well. Yet this is frequently not the case. Many contemporary theories focus exclusively on rules and principles calculated to serve as action guides in an explicit decisional process. Theological traditions of ethics usually expand this to include the virtues, and an emphasis on the non-decisional aspects of moral character. Still we believe theoretical enrichment is essential to be true to the vast scope of activity we routinely designate as ethics.

We invite you to think of the approach to ethical theory we will undertake in this essay not as a linear and progressive rational argument which makes additive and cumulative points for the purpose of unifying or totalizing a theory. Rather we think of our approach as a collage. "Collage" comes from the French *verb* "coller," meaning "to glue." It is typically used as a noun, however, to refer to an artistic production composed of various materials glued to a picture surface. To create a collage is to glue, bind, or hold together diverse fragments and ideas.

To say that ethical theorizing is collage-like is to offer a way to configure ethical thinking in "transition," as Jane Flax says. The objective of the collage image is to configure ethical theory ". . . without resort to linear, teleological, hierarchical, holistic or binary ways of thinking and being" ([5], p. 15). Each theory we will discuss ". . . provides a partial critique and corrective to, the weakness of the others" ([5] p. 15). By putting theories together in a collage-like conversation we will emphasize their collaborative and interactive power.

To say that theorizing is collage-like is to eschew the battle for dominance among theories and to encourage a more complex and less exclusionary approach to moral interpretation. This approach is required by the complexity and multifaceted character of human moral activity and its depth of meaning.

Collages have rough and overlapping edges: cumulative and distinct

171

G.P. McKenny and J.R. Sande (eds.), Theological Analyses of the Clinical Encounter, 171–184.
© 1994 *Kluwer Academic Publishers. Printed in the Netherlands.*

meaning is not what collages picture and not what an analysis of them can yield. Pictures in a collage collect, represent and frame meaning in a simultaneous manner, much like a bodily gesture holds meaning for those who give and receive it. The emphasis is on showing how images are inter-dependent and gain their meaning from the whole, rather than being serially or chronologically anchored.

When we look at a collage often it is at first disconcerting because we are required to set aside the desire to bring one dominant and unifying perspective to bear. Appreciating a collage means keeping the various media in flux, in transition. The more we try to systematize the meaning of a collage from a dominant, single angle of vision, the more the meaning of it is distorted and elusive.

We offer the image of the collage as a way to transform how we con-figure ethical theory. This image is intended to give ethical theory more interpretive power and less explanatory power. It brings to the activity of ethical reflection a shape which is temporally situated, but in which a unitary total-izing perspective is not the goal. The collage image relieves ethical theorizing of the need to push for final conclusions, resolutions about which theory is preeminent or dominant and focuses on interpreting effectively what is before us. It also insists that the experience of ethics-in-use demands multiple theoretical perspectives, multiple tools – which then become part of the activity of excavating ethical encounters.

This essay has two major sections. In Section I two dominant ethical theories, principled reason and what we will call "normative narrative," are discussed. Their theoretical claims will be analyzed and illustrated by reference to a specific case: "Death and Dignity, A Case of Individualized Decision Making," written by Dr. Timothy Quill. A brief rehearsal of this story fragment follows our introduction.

In Section II we will sketch an approach to narrative which emphasizes its rhetorical power and which thereby diverges from the linear and repre-sentational aspects of narrative inherent when stories become normative. This will add new texture to the notion of theory-as-collage we are producing. The rhetorical dimension will allow us to frame features of the Quill story in a new way.

In the conclusion we will summarize the importance of framing the ethical encounter in a collage-like manner by describing what picture of ethical theory emerges when principled reason, normative narrative and rhetorical narrative share a canvas.

Dr. Timothy Quill's Story Fragment

We refer you to Dr. Timothy Quill's story fragment, "Death and Dignity, A Case of Individualized Decision Making," published in *The New England Journal of Medicine*, March 7, 1991 [9]. Here Quill recounts the relationship between he and his patient, Diane, who suffered from acute myelo-monocytic leukemia. For most readers this case is "about" physician-assisted suicide – a discrete set of acts to which the other aspects of Quill's account serve only as backdrop. Yet to see this only as a *case*, and a case of physician-assisted suicide, is already to succumb to a theoretical bias about ethics as explanations of life-or-death *decisions*.

We have chosen Quill's essay to illustrate our points because his account is far more detailed than the typical presentation in which a few sentences are used to serve up an ethical "dilemma." In this sense, the Quill case is more a story fragment, a piece of a longer unwritten narrative about Quill, his life and practice, Diane, her life and family, and their clinical encounters around her illnesses and death. The very eloquence of Quill, the sympathetic characterization of his actions, also presents us with ethical tensions. For while the reader can always probe and question the details of his decision, as was done in the published "Correspondence" in the *Journal* (August 29, 1991), his skill of presentation apparently lulled most readers into complacence about *the mode of presentation* itself. Yet in ethics not only the acts but the representation of the acts have ethical significance. More accurately, we can only know what decisions were made, what acts were undertaken, through the mode of presentation in which they come to us; hence the call for rhetorical analysis.

So, the manner in which we explore Quill's story fragment is *not* designed to explore whether Quill was right or wrong. Indeed, our belief is that that question is only approachable by suspending judgment long enough to see how the questions get framed for us by the theoretical apparatus we use.

I. TWO DOMINANT PICTURES: PRINCIPLED REASONING AND NORMATIVE NARRATIVE

Principled Reasoning

One approach to Quill's essay is to ferret out the principles at work in his decisional process. Doing so yields a way to assess his actions, a way typically employed and recommended in the bioethics literature. For example, Beauchamp and Childress, in their widely-used text, commend principles as a way to bring "order and coherence" to the frequently "disjointed approach" that relies on the discussion of cases ([1], p. vii). The benefits of principles are many. They save moral analysis from an intuitive, ad hoc approach and preserve consistency across a variety of actions and judgments. Recourse to principles also provides a way to justify choices, to say *why* one act is morally preferable to another. An ethical analysis which remained at the level of the concrete and the customary could never provide the sort of force and persuasiveness which complex and contested actions require, and which our reasoning powers demand of us. For all these reasons, a principle-based analysis of the Quill essay is useful. The sketch of such an analysis is given below. Our purpose here is to be illustrative rather than exhaustive.

This analysis will focus on three principles, all of which seem to be satisfied in Quill's account. The first is autonomy, or more precisely, respect for autonomy. Most bioethicists, perhaps most persons in American culture, view autonomy as a paramount value and respect for the autonomy of others as a *sine qua non* of contemporary life. Respect for autonomy means here, first of all, the absence of coercion. A portrait of the interaction between doctor and patient, or family and patient, in which the patient was cajoled, pressured or even gently guided into suicide would be morally opprobrious. Autonomy also encompasses a positive set of values and conditions for an action, for example, the clarity of thought involved in choosing, knowledge about alternatives, having access to unbiased interpretations and whether actions are authentic, that is, consonant with the previous life choices of the actor. A suicide conceived one day and carried out the next would not pass these tests of autonomy, nor would a suicide sponsored by incorrect information or a questionable diagnosis – as may have been the case with Janet Atkins and her physician-assistant, Jack Kevorkian. Though questions can always be raised, Quill's account seems to satisfy many of the requirements of respect for autonomy and its correlative values: "As she took control of her life, she developed a strong sense of independence and confidence" ([9], p. 692). The barbiturate prescription became an extension of this set of values, and was seen by Quill as justifiable, in part, because it enhanced rather than reduced autonomy. Though "uneasy" about his action, Quill reports that he wrote the prescription with a strong feeling "that I was setting her free . . . to maintain her dignity and control on her own terms until her death ([9], p. 693).

A second principle by which Quill's account could be analyzed is beneficence, or positively promoting the good of others and removing them from harm ([1], p. 135). Beneficence is usually understood as a special principle of medical ethics, as well as of general ethics. Doctors are enjoined, thereby, to seek the good of their patients, even when – or especially when – self-interest or the interests of third parties is involved. Medical beneficence is typically understood not simply as justified and laudatory, but as required, that is, a *duty* of physicians. In this case, Quill has a duty, under the principle of beneficence, to actively promote the good of his patient Diane, and to remove harms. Here again it would seem that Quill's story yields a justifiable picture of his choices. One of Quill's aims in his actions was enhancing patient self-determination and here autonomy is the guiding principle. Another aim was the prevention of suffering, the avoidance of an agonizing and prolonged death. Here beneficence is the principle at play. Such a possible harm was, in this case, exacerbated by Diane's particular fear of an extended and painful final course. The prevention motif comes into play again after Diane's death in Quill's choice of how to list the *cause* of death. Here beneficence (for Diane's memory, for the family, for himself) overrode any exacting requirement for truthfulness. "When asked about the cause of death, I said 'acute leukemia'. . . . I said 'acute leukemia' to protect all of us. . . ." ([9], p. 694].

In establishing a justification grounded in beneficence a great deal depends on interpreting (or believing) self-reports of motivations and intentions. In this case the beneficence seems to shape the pattern of interaction over the entire course of the patient-physician interaction and so becomes believable when self-reported by Quill as a motive for the critical choices. Falsification of a death certificate solely in the service of hiding complicity in Diane's action would not have qualified as an instance of beneficence.

A third principle is less easily encapsulated. It involves due regard for the well-being of others besides the patient, and in this sense, could be called an extension of beneficence. Yet it is more than just their well-being which is at stake. That more involves the essential social relatedness of persons, the recognition that none of us are isolated individuals, or moral atoms – that all our acts affect others, sometimes profoundly. Autonomy and beneficence, though they may be paramount, cannot then be the only principles. We call this third criterion *justice*, to indicate the need to recognize multiple persons and interests at work. Since Aristotle "justice" has been roughly formulated as giving each person his or her due, that is, treating equals equally and unequals according to their relevant inequalities. What that means in this case is determining the proper role, voice and influence of Diane's husband and son. Their acceptance (though not necessarily concurrence) in Diane's decision adds strength to Quill's claim for a justified action. While spousal or family interests to the contrary may not override a rational choice for suicide, their opposition to it, or their being wronged or damaged by it, would have weight. In such circumstances the requirements of justice could not be (completely) satisfied. Such inability to meet fully the principle of justice may not prove ultimately persuasive, for one could argue that justice is not the prime principle here. And it is a rare circumstance to find a moral resolution that will satisfy all or even most of the ethical principles at work. Still, the fact that the rights and well-being of the family were taken into account, and that the family was seemingly brought closer together in the process, is another reason to think that the right thing was done.

It is useful to note here that Quill never makes reference to principles in his account. To argue that principles are the primary mode of moral understanding is, therefore, to assume that the account opens itself to principled analysis in some fundamental way, as if the reader should follow the account and insert the principles at the right spots.

We would not claim that the interpretation of this case has been exhausted by these principles, or even that we have applied them in an uncontested way. Rather we have aimed to portray how a typical, principle-based analysis would look. This portrayal indicates both the strengths of principles, but also why "narrative" is often taken as an alternative interpretive form.

Normative Narrative

Over the past two decades narratives and story-telling have frequently been offered as forms of moral interpretation which are superior to principled reasoning. Narratives, on this model, are designed to provide the norms of interpretation. Just what this means is best explored by beginning generally, with narrative as a category of experience, and then drawing forth the implications of narrative and story for ethics.

Stephen Crites' essay "The Narrative Quality of Experience" is a good place to start. Crites' claim is that "the formal quality of experience through time is inherently narrative" ([3], p. 291). Crites' thesis is Kantian, for what concerns him is not the persistence and ubiquity of stories, but the *necessarily* narrative structure of any and all human experience. Quite simply, our experience requires a narrative quality as a condition of being what it is. This narrative quality is not something we choose to have *in* our experience, but a condition of their being experiences at all. Narrative is, in short, an *a priori* inner form rather than an interpretive category.

In one sense Crites' claim seems self-evident. There is no present without a past and a future. Human existence is "tensed" and would be incoherent without "the tensed unity" of "these three modalities in every moment of experience" ([3], p. 301). The thesis is as old as Augustine who laid it out definitively in *The Confessions*. In another sense this is a grandiose claim. The assertion that narrative form is logically and existentially primitive, that there is no way to get distance from it or get critical reflexivity on it as a category of experience, is immense. In narrative, Crites says, we reach a kind of "bedrock," and in Wittgenstein's sense our "spade is turned" ([11], p. 85e).

It is difficult to see how such a claim could be proved or disproved, but that does not concern us here. Rather our aim is to see the implications of this claim for ethics. Crites begins to draw forth the implications when he concludes that "ethical authority . . . is always a function of a common narrative coherence of life" ([3], p. 310). But this is only a beginning, for if narrative is experiential "bedrock," a profound reorientation to the field of ethics is required. Two of the most influential figures is formulating this reorientation over the past decade have been Stanley Hauerwas and Alastair MacIntyre. Hauerwas speaks explicitly out of a theological grounding, while MacIntyre's work is rooted in an approach to the history of philosophy. Yet they share a deep dissatisfaction with ethics as principled analysis and both see narrative as fundamental to a more sound approach. They present logical extensions of Crites' thesis for the field of ethics.

Hauerwas' sense of the primacy of narrative is perhaps best stated in his essay with David Burrell "From System to Story: An Alternative Pattern for Rationality in Ethics" [6]. As the title states, the aim of this essay is to set forth the narrative dependence of all "systems" of ethics and to argue for a form of rationality in which the coherence of truthful and empowering stories displaces the logic of formal reasoning. Religious convictions are seen as "canonical" stories, which give fundamental orientation and cohesiveness to moral selves. The problem with principles arises, in Hauerwas' view, with the assumption that they are somehow outside of any and all stories, that they are "narrative-free." This wrongly places them in a superordinate position over narrative interpretations of morality. More specifically, the principled approach sees stories as hortatory homilies, or mere illustrations of principles. Narratives are to be judged, then, and the issues they depict adjudicated, in terms of principles. Narratives can be vehicles of moral insight, but only vehicles. There is nothing intrinsically valuable about story-form and every story must be weighted in the scales of principled analysis to order to harvest the moral assessment of the actions it contains. The narrative container is at best neutral, and at worst a distraction or impediment to the real business of ethics.

In contrast to this, Hauerwas believes that "character and moral notions only take on meaning in a narrative" ([6], p. 15). In essence, every "system" of ethics is at root a story, a narrative of character, virtues and selves displayed in some normative way. Principles, instead of being the norms *for* stories, are actually short-hand reminders for these larger narratives of the self. In Hauerwas' view we understand moral principles rightly when we see them as grounded in narrative and presenting in truncated form the canons of some story. The place of principles in our moral judgments, therefore, depends on the interpretive truthfulness and power of the narrative which they represent. Ultimately it is not principles but stories which compete for interpretive primacy. In brief, it is not just that dealing with principles is dealing with abstractions. It is also that principles, in this view, represent surface norms which we are likely to use poorly or incorrectly unless we track them to the root narrative they stand for.

MacIntyre's view is consonant, but his emphasis different and his aim more broadly cultural and philosophical. *After Virtue* is a sustained argument for an Aristotelian, teleological view of ethics in which the virtues can be reinstated as the chief component of moral self-understanding [8]. As part of this, MacIntyre also reinvigorates narrative as an essential concept and story-telling as an essential activity. For example, in his critique of Sartre's episodic, existential concept of action, MacIntyre argues that "the characterization of actions allegedly prior to any narrative form being imposed upon them will always turn out to be the presentation of what are plainly the disjointed parts of some possible narrative" ([8], p. 200). He concludes: "Action is always an episode in a possible history" ([8], p. 201). As for story-telling, MacIntyre sees it as an activity central to the human situation and a necessity for moral self-awareness. "I can only answer the question 'What am I to do?' if I can answer the prior question 'of what story or stories do I find myself a part?'" ([8], p. 201). For MacIntyre the end-point of his arguments is not the acceptance of canonical stories but recognition of the bankruptcy of what he terms the "Enlightment project," that is, the effort to interpret moral life solely by recourse to independent, timeless principles. In place of these principles MacIntyre would have us espouse a teleologically grounded (narrative) notion of the virtues and a coherence of life given unity by the practice of the virtues.

Again, we do not wish to argue for either Hauerwas or MacIntyre here, but to indicate how these theories taken together, provide a reorientation of our notion of ethics. But what difference would this make in interpreting the Quill case?

*   *   *   *   *

To see Quill's account as fundamentally narrative is to ask about the character of the agents involved. For example, working from narrative, we might ask "Is this story a truthful or self-deceptive account?," "What virtues are displayed?" and "In what larger story is Quill's narrative rooted?" Thus, seeing ethics as embodied, necessarily embodied, in narrative form brings different kinds of question to the front. The sort of moral coherence it seeks is not that of reason, but the coherence of characters and histories with whom it is possible to identify and lend our concurrence and admiration, or our suspicion and disdain.

The limitations and liabilities of principled approaches are the assets of narrative approaches. Indeed, the Quill presentation seems written as if inviting a narrative-informed analysis. It is told *as a story* – his story. The title itself "A Case of Individualized Decision Making" seems to warn the reader away from generalization and toward appreciating the particulars, or unique qualities of his experience. Yet narrative approaches are not free from their own hazards.

One hazard of a narrative approach is the tendency to acritically adopt the point of view of the narrator as normative for the story presented. Quill, for example, is skilled enough in his portrayal to elicit our early sympathy for him and his patient. Too much sympathy can blunt our critical faculties and encourage premature closure on the ethical conflicts. In this sense, the ease with which Quill's account can be adapted to narrative theory may signal a deeper problem beyond the surface fit.

Another hazard is the discouragement of multiple perspectives implied in a single narrator. If there is one narrative account of this set of clinical encounters, could there not be several? A powerful narrative can serve to block alternatives, multiple interpretations, with different authors and different renderings of the "same" events. A powerful narrative may serve only to reinforce our prejudices, especially if we identify with the narrator. It may be true that, in some sense, moral notions of the self are figured in a narrative. Yet care is required that being figured in a narrative does not result in being grounded in narrative as an absolute or privileged model. We should resist the move from appreciating a particular story to "Story" as the model for all facets of the ethical encounter.

A third and less pedestrian problem has to do with the relentless, diachronic temporality which can accompany narrativity. Beginnings, middles and endings – neatly tied into a temporal sequencing – can be as tyrannical over the moral imagination as the logic of principled reasoning. Deductive systems of principles tend to trap moral agents within the exercise of logical maneuverings, as if agents were in an administrative posture over themselves and their actions. Narrative accounts of ethics correct for this deductive bias but tend to confine the agent within a sequential, linear course. In narrative theories the tyranny of principles is avoided but only to be replaced by the threat of a tyranny of teleology. Narrative forms threaten to turn the vertical logic of principles into the horizontal logic of story. One form can be just as exclusionary and absolute as the other.

II. FINDING NEW PICTURE FRAMES: RHETORICAL INTERPRETATION

Narratives as normative for agents and actions, as canonical stories, is only one way of formulating the power of narrative interpretation. Narrative theory has many different voices: Russian formalist theories, dialogical theories, New Critical theories, Chicago school, or new-Aristotelian theories, psychoanalytic theories, reader-response theories, and post-structuralist and deconstructionist theories.

It is not our intention to outline or assess these various voices. Rather, with Edward Said, we want to ask how the narrative "provides representational or representative norms selected from among many possibilities. Thus the novel (story or narrative) acts to include, state, affirm, normalize, and naturalize some things, values and ideas but not others, yet none of these can be seen directly" ([10], p. 179). It is the hidden exclusion or selectivity of normative narratives that we want to get at by a rhetoric of interpretation.

Speaking of the novel as a particular form of narrative, Said goes on to say that formal theories "make sure that the novel's remarkably precise articulation of its own selectivity appears simply either as a fact of nature or as a given ontological formalism, and not as the result of sociocultural process. . . . Apologists for the novel continue to assert the novel's accuracy, freedom of representation, and such, the implication of this is that the culture's opportunities for expression are unlimited. . . ." ([10], pp. 176–177).

It is our belief that formal theories, in ethics, as well as in literature, push for universal accuracy, for representation unfettered with contingent purposes, and hide their own methods of selectivity. Making explicit these hidden methods of selectivity is a task of rhetoric. Frank Lentricchia, working in furrows plowed by Kenneth Burke, offers these rhetorical tools of analysis. In *Criticism and Social Change*, Lentricchia claims that Burke wants

to revise the traditional idea of rhetoric so that, newly conceived, it could stand between pure deliberate activity and the unconscious. "There is an intermediate area of expression that is not wholly deliberate, yet not wholly unconscious. It lies midway between aimless utterance and speech directly purposive." "Speech directly purposive" stands, of course as a definition of rhetoric in its classical, overtly political project, while "aimless utterance" signifies what the literary has become from Kant to de Man: discourse presumably apolitical, unrhetorical, without a design. . . . Burkean rhetoric would occupy the space between the old rhetoric of pure will and the modernist and postmodernist aesthetic of antiwill: between a subject apparently in full possession of itself, and in full intentional control of its expression, and a subject whose relation to "its" expression is very problematic ([7], pp. 159–160).

Such a picture of the subject in full possession of itself is the companion to a notion of reason as the universal and apolitical arbiter of that self in society.

A rhetorical approach gives to narrative interpretation the capacity to interpret ethical acts and ethical scenes as activities of a subject positioned somewhere between the old rhetoric of pure will and the post-modern position on subjectivity. Because the post-modern position seeks linguistic meaning through signifiers on a page, the agency of the subject is effaced. A rhetorical approach does not efface the agent, but it does make the agent's place contingent. This is because the *field of action* and its mode of representation become as important, if not more important, than the actor. The field of action is not suspended outside the contexts and circumstances of its actual social-political web of meaning. Because the field of action is taken seriously, the agent has less control and is affected by the meanings of the political-social world. The agent is now seen in relationships and these relationships, rather than the intention of the narrator, become the web of meaning in which agents are portrayed and out of which they speak. To pay attention to the rhetorical aspects is to introduce contingency, for rhetoric is not about picturing meaning in an isomorphic representational model, but about persuasion. As Burke says, "wherever there is persuasion, there is rhetoric. And wherever there is 'meaning', there is 'persuasion'" ([2], p. 172).

The sense of self or actor that dominated much of normative narrative theory (and against which the rhetoric of the postmodern interpretive bent is directed) goes something like this: Narrative assumes an essential correspondence between the autonomous subject, his or her intentions, and personal identity. Against this dominant interpretive whole,

Burke brings his central proposal of "action" not as the production an "agent" who would be in turn conceived as the expressive origin and master of "action"; but as a transpersonal, all-embracing process within which agents are always already situated. . . . Burke argues that cultural trends emphasizing "all things are full of Gods" (as Thales thought) would translate, in our more resolutely secularizing vocabularies, into "all things" are full of "powers" or "motives" that cannot be grounded in agent terms like "personal identity." *The collapsing of conceptions of motives, powers, and action into the subject-agent is an effect of what Burke calls a "capitalist psychosis"* – the rise of private property as a privileged category of value and the concomitant reconstruction of identity, motives, action, and power as analogues of private property, "possessions" of the subject ([7], p. 139, emphasis added).

Normative narrative has frequently served as a tool in ethics to expose the inadequacy of rules or principles to express a full range of human meaning in a given ethical encounter. It is the "pure will" signification of rule or principle-dominated ethical deliberation that narrative has sought to overcome. However, a concomitant revision of self, along the lines outlined above, has not taken place. Proponents of narrative as normative have considered it to offer a more adequate understanding of the social situatedness of the agents in the clinical encounter. But that encounter has not been critically located within a capitalist culture and "agency" has not been critically reformulated to acknowledge its "capitalist psychosis" tendencies.

A rhetorical understanding of narrative situates the specific conversation of ethical deliberation in a larger historical-social-cultural history. This is necessary because the narratives are not stories about people who already "possess" their identities. The individual stories that make up a specific ethical encounter are already situated in a social and political field of action which can *not* be suspended when an ethical encounter happens. Ethical stories are part of an ongoing conversation between and among medicine/science/culture (doctors and patients) in which each specific encounter in the clinic is no "Once upon a time." So, as Lentricchia says about Burke's primal scene of rhetoric, "we can't know what we would ideally like to know about the scene we've entered because no one was there at the beginning, when the conversation started" ([7], p. 161). Reason cannot be the universal arbiter of this kind of multiply diverse and contingent history nor can normative narrative theory presume to contain it in story form. A rhetorically grounded use of narrative says we find ourselves in conversations that we make, but that are always bigger than any one person or any one story. We feel burdened immensely because we are not the master. But, then we act. We decide with no formal assurances; ultimately the ethical encounter may be an occasion for agreement or disagreement, or just confusion, a deliberation without strict rules, a story in which we cast our oar and enter it, without assurances. As Lentricchia puts it: "This is a conversation without epistemological 'foundation' or 'substance'. The fact is that we are going to argue with one another, heatedly, and the argument is not going to come to a conclusion (no 'Happily ever after') because more people, in similar states of historically burdened ignorance are going to come into the room" ([7], p. 161).

The making and being-made rhetorical structure of cultural situatedness is always *already* underway when any particular ethical encounter is discussed. Founding ethical theory on being able to suspend this historical reality has required epistemological clarity and ontological purity when such is not possible.

Agency Reconfigured

Rhetorical interpretation allows us to put different texture into the theoretical collage. It asks different questions of how and why stories mean what they mean. It acknowledges, as Terry Eagleton says, that "meanings are products of language, which always has something slippery about it. It is difficult to know what it would be to have a 'pure' intention or express a 'pure' meaning" ([4], p. 69). It is also difficult to know what it would be to have "full" possession of oneself and have "full" intentional control of my expression.

In both principled reason and normative narrative, agency – the subject's production of ordered actions – relies on an understanding of how decisions mean or stories mean where meaning is capable of being fully separated from the rhetorical field of action. Both theories assume, in one fashion or another, that we can know the ethical encounter as it is, and that we can do this because the agent is the sole producer of its meaning, or that its meaning can be made transparent in a canonical story.

When we ask questions about *who* the subject is, we are asking questions about the subject as actor. They revolve around a conception of a subject as the active or productive mind. In principled reason we have a picture of reason as the universal and apolitical arbiter of agent in society. This view of reason's production of meaning permeates and defines the self-portrait: a self in full possession of itself and in full intentional control of its expression. The language of this agent looks much like the old "rhetoric of pure will."

In normative narrative we have less a picture of the subject or author in full possession of pure intention, but a picture of the subject completely knowable, or transparent, to the terms of a canonical narrative. Here the agent's identity is not produced through reason, but discoverable through a story. Yet both these dominant pictures present a picture of subject and meaning, a picture of actor and action of the story which is essentialist. Hence, questions of the already situated nature of the ethical field of action remain unattended. Questions of agency, then, take the following form: *WHO* is the *SUBJECT* of the story?

In order to shift our attention from the familiar questions of agent-as-subject to the rhetorical dimension we must ask the following question: *WHO* is the *OBJECT* of the story? This is not a question about the narrator, the actor, the agent, the active mind, the producer of the story, or about the hero, protagonist, or character embedded *in* the story. This question moves agency from the central position it always occupies in either of our ethical theories, e.g., as decision maker, author, or as protagonist and lead character. The agent is, consequently, no longer regarded as the sole source, or the ultimate telos of meaning.

Who is the object of Quill's story? We are after a rhetorical reading that places emphasis on a subject objectively related to a context outside the ethical encounter itself. This reading turns the question of agency into a question of the social-political field in which agents act. This field is always one of the ethical encounter *plus* its larger frame of reference. Both anchor the way ethical dilemmas are represented.

By asking who is the object of the story, we are enabled to disattend from questions of the agent's active production of meaning. This rhetorical reorientation enables us to move away from an exhaustively individual-productive reading of agency. Individual subjects are always actors in a field of action. This site of rhetorical activity is created when we ask about wider relations between Quill's story and his purposes, audiences and constituencies which lie outside the particular story he narrates.

Put another way, the rhetorical activity of the story reveals a hidden *who*, a *who* which is the *object* the reader can see in the story, not created by the story, and not the subject of the story. In this rhetorical picture, intentional agency does not have to bear all the weight of the ethical gaze. It is the rhetorical appeal which takes center stage. How the story represents its motives and *who* it appeals to – its audience, constituency and destination – becomes a vital part of its ethical dimension.

We will not attempt a full rhetorical interpretation here, but simply indicate how one might begin a rhetorical analysis of the Quill account. Note, for example, that this is a story which begins with "I's" and descriptions of the patient, Diane, and develops into a "we" at key points in the story. Diane requests barbiturates and it is the "I" of Quill who deliberates, is troubled, yet prescribes them. It is individuals who respond and form an interaction, but it is a "we" who hope for a remission and a "we" who "had no miracle." After Diane's death a new "we" is formed as Quill talks with the surviving husband and son. The professionally beneficent "I" re-emerges to report the cause of death to the coroner. This "I" persists to the end of the story fragment, but in the last three paragraphs the "I" as interpreter takes explicit shape and seeks a "we" with his readers. Here the discourse ceases to be descriptive and becomes hortatory, contemplative and finally, metaphysically specula- tive. Quill becomes less a narrator of events held up to public scrutiny and more a person reaching out for an interpretive bonding with a professional audience of physicians, and perhaps also the community of the terminally ill and their families.

To trace "I's" and "we's" is to ask about fluctuations in agency, and to see – not a simple and discrete subject – but a subject in a field of action where relationships oscillate. Those in relationships, by definition, cannot be in clear and simple possession of the moral significance of their actions. Being in relationships mean one's moral significations are now irretrievably mediated through these. Rhetorical readings are then another disproof of the moral sovereignty of the atomic "I," so much the focus of contemporary ethical analysis. To read for relationships rather than acts and reasons, or characters and virtues, is one aspect of what we mean by de-centering the subject and seeking the *who as object* of the narrative. It moves the analysis from a focus on self possession/expression to the field of relationships which hold the agent both as subject and as object. Generally, we believe it is more helpful to see Quill as building a moral collage of interpretive strategies (and inviting the readers' attention to these), than to see him as engaged in deductive justificatory schemes, or in rehearsing a story which will fit a canonical pattern.

$$* \quad * \quad * \quad * \quad *$$

Principled reasoning and normative narratives are powerful and important tools in ethics. We do not argue that they are false, or useless, but that it is mistaken to see them as all-embracing alternatives which must subsequently compete with each other. Competition leads to assumptions about dominance and subservience, winners and losers. To see these differing approaches as antagonists in pursuit of the status of master theory in a winner-take-all contest is to see them as still-shots competing for representational control. The collage metaphor is intended to thwart our expectation that a single reality is "out there" to be caught in some mirroring still-shot. Theories in ethics are not like photographic images, to be judged by monodimensional or isomorphic "fit" to the world. What is there to be grasped (always incompletely) in a theory is not a simple, or stable moral reality, but agents and relationships in complex webs of social and political, as well as strictly moral meaning. Theory which even approaches adequacy must be layered and textured – like a collage – and capable of interpretation from a wide variety of perspectives.

The broadest frame for the collage is this: the meaning of ethical theories lies in their use in the practice of living lives which are themselves collage-like. The usefulness of rhetorical analysis in interpreting this collage lies in its reflexivity – its power to turn ethical discourse back upon itself, frequently at cross-purposes with principled reasoning and story-telling. Rhetorical analysis embodies, then, a view of moral interpretation as chiastic – crisscrossing the vertical lines of reason and the horizontal planes of stories.

On the collage model, the power of ethical theory resides *not* in how well it founds or justifies the claims of the system – principles' reasons or narratives' stories. Rather the power of ethical theory resides in its capacity to frame the ethical encounter from as many perspectives as necessary to keep footing and friction – for thinking and judging wisely.

BIBLIOGRAPHY

1. Beauchamp, T., and Childress, J.: 1979, *Principles of Biomedical Ethics*, 1st ed., Oxford University Press, New York.
2. Burke, K.: 1969, *A Rhetoric of Motives*, University of California Press, Berkeley.
3. Crites, S.: 1971, 'The Narrative Quality of Experience', *Journal of the American Academy of Religion* XXXIX, 3 (September), 291–311.
4. Eagleton, T.: 1983, *Literary Theory. An Introduction*, University of Minnesota Press, Minneapolis.
5. Flax, J.: 1990, *Thinking Fragments: Psychoanalysis, Feminism, & Postmodernism in the Contemporary West*, University of California Press, Berkeley.
6. Hauerwas, S.: 1977, *Truthfulness and Tragedy: Further Investigations in Christian Ethics*, University of Notre Dame Press, Notre Dame, Indiana.
7. Lentricchia, F.: 1985, *Criticism and Social Change*, University of Chicago Press, Chicago.
8. MacIntyre, A.: 1981, *After Virtue*, University of Notre Dame Press, Notre Dame, Indiana.
9. Quill, T.: 1991, 'Death and Dignity, A Case of Individualized Decision Making', *The New England Journal of Medicine*, vol. 324, 10, 691–694.
10. Said, E.: 1983, *The World, The Text and the Critic*, Harvard University Press, Cambridge.
11. Wittgenstein, L.: 1953, *Philosophical Investigations*, G.E.M. Anscombe (trans.), The MacMillan Company, New York.

RICHARD M. ZANER

ILLNESS AND THE OTHER

1. CLINICAL PRESENTATION OF THE OTHER[1]

Not long ago I received a call from a physician who wanted me to stop by to see one of his patients – a young man who had been hospitalized a week prior and was adamantly refusing hemodialysis. Although dialysis would surely be beneficial, his refusal would mean he would die.

In his late '20's, he had been hospitalized numerous times over the previous summer for all manner of problems and was apparently just fed up with everything. Reading his chart I was stunned. This young man's life seemed to thread out from an infancy I had encountered many times before with premature babies in the Newborn Intensive Care Unit (NICU). He had been born with spina bifida: spinal column protruding from a large lesion (myelomeningocele) high on the column, he was paralyzed from that point down making him incontinent, paraplegic, and hydrocephalic. He had a surgically implanted shunt that took the cerebral-spinal fluid from his brain to his abdomen and, like any such child who survives and manages to grow, had experienced repeated surgeries over the years to replace the shunt.

That summer he began to have severe diarrhea, dehydration, infections, and a malfunctioning bladder – accounting for the many hospital admissions. As if all that were not enough, the renal failure that had become apparent in the last month or so was thought to be irreversible; the diarrhea and infections continued; and now he had become anemic. On the admissions form it was indicated that he had refused dialysis at another hospital and this was given as the reason for his transfer to the present hospital – although, as mentioned, dialysis would likely enable him to return to home and possibly to the job he had held for some years. Thinking that the young man was depressed, his physician prescribed an anti-depressant in the hope that, in a less distressed state, he might change his mind.

His mother had accompanied him to the hospital. She expected that he would be placed on dialysis. Perhaps, she thought, he might even at some point be placed on the waiting list for a kidney transplant. But he was no more admitted than he began refusing dialysis. Fortunately for those in charge of his care he didn't need it immediately so it was "easy," as his attending physician told me, to postpone things and start the anti-depressant. Then, when he changed his mind – as was fully expected – they could snake a catheter into his arm and he could be hooked up to the machine. The toxins that his kidneys could no longer manage would be mechanically removed and he could then be evaluated for transplantation.

185

G.P. McKenny and J.R. Sande (eds.), Theological Analyses of the Clinical Encounter, 185–201.
© 1994 *Kluwer Academic Publishers. Printed in the Netherlands.*

Despite the anti-depressant, however, he continued to refuse treatment. Within a few days the poisons inevitably accumulated and dialysis was imperative. His mother was terrified. His physician had to go out of town for the weekend and the resident and nurses seemed at loggerheads with his persistent refusal. In the press of circumstances the physician covering for the attending decided to have a psychiatrist assess competence. Surely, it was thought, no one in his right mind would refuse treatments that promised relief and possibly even a return to his normal life.

Not surprisingly, by that time the psychiatrist found him "temporarily incompetent"; his "decision against dialysis" was then taken to be a function of depression and failing thought-processes – which were themselves said to be caused by the build-up of poisons in his bloodstream. Armed with that, the covering physician placed the catheter and took him off to be dialyzed.

When the attending returned the next Monday he was concerned – furious is probably more accurate. Not only had his patient been dialyzed despite refusing it over the past week, but here the young man was, alert once again (treatment had, of course, done its thing) and very disturbed at being forced to have dialysis. In fact, he continued his adamant refusal of any further sessions. Both the attending and his mother were in a bind. Both wanted to respect what they considered to be his competently expressed refusal. On the other hand, knowing that dialysis was actually beneficial – they had living proof of it before their very eyes! – they also felt the crunch of that knowledge. Yet, he repeated that he had been through enough; in fact, had a lifetime of enough! What to do? Well, call in the ethicist!

When I agreed to the consult, thoughts of all those babies I had seen in the NICU were spinning through my mind. It was as if one of them had suddenly grown up and was now demanding to know why he had been forced to stay alive. Until now, I had been able merely to wonder: what did the future hold for those many babies? Well, here it was, in living color. Meet Mr. Thomas Pembroke Brown, my very own apprehension come to life. Except, like most worst-fears-come-to-life, he proved to be very little like what I had until then only imagined. Thus does reality make fools of us all.

When I went into his room his mother was sitting next to his bed. Mr. Brown was lying with his eyes closed, as if asleep. I said hello to his mother and introduced myself. Before I could say anything else he opened his eyes and said "Hello, I'm Tom," and the barriers (those spina bifida babies) that my overworked imagination had unwittingly constructed collapsed. He looked and sounded *normal*.

Discussion with the attending and reading Tom's chart led me to focus on issues about death. It seemed to me that Tom and his mother both needed to think and talk about that and to do so together. I wondered whether they had actually done much thinking and talking about what would inevitably happen should he not be treated. Did he know, as his mother surely did, that refusing treatment would bring about *his* death?

After some small talk, I turned to Tom with the intention of opening that sensitive topic. Suddenly his eyes rolled up, he began gagging and twitching, and his faced contorted horribly. *My lord, . . . what've I done? What's happening?* I quickly called in his nurse; she calmly took charge, informing me along the way that Tom was having another seizure. This was clearly not a time to start up a conversation about death and dying. Keeping some presence of mind, I took my leave. I expressed my concern and apologies to his mother and promised to come back the next day.

I was really shaken as I walked away from his room. This was the first time I had witnessed a *grand mal* seizure in person and I was deeply troubled. My thoughts and feelings were turbulent. I even found myself foolishly wondering whether I had brought it on. As I later found out when I talked to his doctor, everyone had become so centered on Tom's refusal of dialysis that they had forgotten to tell me that, after years of being effectively controlled, his seizures had recently returned. As if everything else that he had endured over the past few months were not enough, now he had to contend with that. Thinking about that I must say that his refusal of dialysis began to take on greater weight. Realizing, however, that one's natural tendencies can be naturally misleading, I worked to mute that inclination.

The next morning I came back to see Tom and his mother. Things were much calmer and we moved directly to the issues at hand. Did he understand the implications of his refusal of dialysis? In a way, he did; but as we talked it seemed to me that he hadn't thought it out at all well. He was in fact behaving rather differently than one would expect when meeting someone firmly refusing potentially life-saving treatment. It's not that he was calmly acceptive. He had not in fact discussed the matter with his mother, had not really even thought about it much for himself. He had not signed any advanced directive – the idea hadn't even occurred to him. Nor had his mother raised the issue with him, neither the advanced directive nor the clear consequence of his refusal. The thought that he would die without dialysis had, so to speak, sort of sidled past his awareness now and then, but he had not confronted matters squarely; nor had she made her own feelings explicit with him.

Trying to learn more about him personally, I asked him about his job – as it turned out, one which he obviously enjoyed enormously and in which he took considerable pride. He had a position in the main office of a state agency, one that gave him a good deal of independence. He had lived with his mother since birth and she had taken care of him that entire time. Before he had become so sick some months ago, he had begun to think that he might, finally, get an apartment and begin to live on his own. *That*, it dawned on me, was what was really on his mind. The numerous illnesses and hospitalizations had eventually required him to quit his job. This, more than anything else, seemed the source of what had been labeled his "depression." Like any of us at one time or another, he had then "figured it all out." On dialysis he would not be able to hold a job, much less go back to the one he really liked. *Ergo*, life just isn't worth it, so let's just give it up! Being *normal* – that is, working and living independently – had become an insurmountable goal when viewed from his perspective.

He continued talking and I listened. A few months ago, he said – almost as an afterthought – he had been told by his supervisor that his job would be waiting for him when he was able to return. As soon as he said this he noticeably perked up; his talk became more lively; his gestures more animated. "That's right," his mother quickly affirmed, "Mrs. Y did say you could have your job back when you're able."

"But how can I work," Tom's tremulous words seemed at once hopeful and wary, "when I've got to be on that damned machine so much?" His mother and I vied with each other to get the thing said: work was indeed possible. Hadn't he discussed this with his doctor? He wasn't sure. Perhaps he had been so wrapped up in grief and a deep sense of loss that he hadn't heard; perhaps none of his doctors had thought to mention it – or, if any of them had, Tom hadn't understood.

In any event, it was by now clear that the way things appeared had changed dramatically for him. I suggested that he really needed to find out much about dialysis and to call up his supervisor to check with her about returning to work. The conversation then wound down. It was perfectly obvious that he did not want to refuse dialysis and that he desperately wanted to get out of the hospital and back to work. For that it was necessary to set up meetings with his doctor who would be able to help him arrange a convenient place for outpatient treatment – and, eventually, be evaluated for a transplant.

Before things concluded, I felt the need to bring some resolution to the issue that had brought me into his life in the first place: his refusal of treatment. I wanted to emphasize how important it was for him to confront that issue. Both he and his mother needed candid and continuous talk about death – especially the need for him to set out his wishes clearly. After all, he was still not well. He remained seriously ill. He might in fact find himself once again in the hospital and, perhaps, unable at that time to voice his wishes. It would be unfortunate were that to happen and for his mother to find herself in a situation in which she did not know what he would want done or not done.

We talked about the matter for a bit, including some discussion of the new state law on advance directives, do-not-resuscitate orders, withholding and withdrawing of life-supports – the whole gamut of issues his medical condition might well bring about. I mentioned, too, how deeply affected I had been by those tiny babies in the NICU – especially what happened to them after they were discharged – and shared my apprehensions about first meeting him. Then I left, thanking him and his mother for allowing me into their lives under such difficult circumstances and emphasizing how much they had helped me understand.

As I was walking out, he called out, "Hey, doc, what do you *really* do for a living? How do you make your bucks?"

"I guess you could say that I'm a teacher, a college prof, Tom. Only, my students are medical students, young doctors, attending physicians, nurses – even patients and families, people like you and your mother. . . ."

"Well," he quipped, "if you ever want me to talk about these things with those *other* students of yours – remember? You asked me about this yesterday. Well, just let me know!"

Had I done that? I wondered. I couldn't recall – I'd probably been too stunned by events and his seizure. In any event, I told him I surely would and would also check back on him periodically.

2. THINKING BACK ON THE CLINICAL ENCOUNTER

This encounter was a first for me, as I had not hitherto encountered a patient like this: alert, competent, refusing treatment that was in some sense beneficial.[2] Tom was quite obviously able to decide things for himself. He had in fact done so several times. That he was "depressed" was probably correct – although it was just as probably doubtful that it was psychiatric in nature (permitting that dubious override of his wishes). That he needed to be confronted with the nature and consequences of his decision was doubtless also true. It could well be, I suppose, that his physician's assumption that Tom was *not* really informed about the matter – and that he therefore needed me to tell him about things – was simply an expression of the doctor's anxiety, even denial. That may be too harsh, for his physician is a deeply caring person. Faced with what seemed by any reckoning to promise benefit, what was he to think about that refusal: did Tom truly understand what he was doing? Did the physician?

I remain unsure. Questions continue to haunt me. When Tom first expressed his refusal to undergo dialysis, shouldn't that have had priority? Did his decline into renal psychosis change the competency with which he had in a way chosen that result? The decline, after all, was exactly what everyone should have expected to occur. Within a day or so more he would have lapsed into an irreversible coma and then died. Wasn't this just what he had competently chosen?

By the time I got involved he had already had one course of dialysis, had markedly improved, and was still adamantly expressing his refusal. Shouldn't *this* decision have prevailed? For that matter, what business did I have going in there and talking about death, living wills, durable powers of attorney for health care, do-not-resuscitate orders, and the like? However helpful I was trying to be, wasn't that just another form of coercion?

Thinking about that brings to mind something else. If it was appropriate to call in a psychiatrist on the weekend, wasn't it just as appropriate to do so earlier? Especially if some sort of depression was suspected (and medication given)? What would have been the assessment at that earlier time? If it was proper to call in the psychiatrist on the weekend, moreover, wasn't it also appropriate to have him evaluate Tom on Monday when he again refused dialysis? If so, why ask for an ethics consultation? In fact, were there any "ethical issues" at all?

Consider an alternative: was the psychiatrist needed at any point? Living as we do under the umbrella of autonomy, if Tom was competent to make his own decisions early in the week, then there seems no question but that it was his "right" to make whatever decisions he wanted – and that his doctors and nurses had no choice but to comply. Shouldn't that have closed the issue? Treatment is offered and it's refused: once a competent person chooses, the game is over for the rest of us – whatever we ourselves may or may not think we would do in similar circumstances, and whether or not we like what Tom decided. By what right, then, could anyone else gainsay Tom's choice, his exercise of self-determination?

But then there is his mother. He didn't seem concerned with what his decision was doing to her, or would do to her if he had his way. Perhaps that should have been the theme of my discussions with him. Perhaps I should have been asked to talk simply with her. Perhaps talking about death did provoke something in him, for he did begin to think about things he hadn't considered before: his mother, job, relations with other people, and the life he really wanted but couldn't see well enough through such dark lenses of grief and so many trips to the hospital.

3. RE-ENVISIONING THE ENCOUNTER

Tom, lying in the hospital bed, his mother seated next to him. Her hand lying on the cover next to his arm, his eyes closed. She looks up at my entrance into the room, her eyes questioning. I have the sense that I'm intruding, but her eyes draw me further into the room. She wants to know what I'm there for. I say "Hello" and she returns the greeting. Then I tell her my name and that Tom's physician asked me to stop by to talk with them. I tell her I'm in "ethics." Her eyes don't tell me anything, whether she heard the word or not, or whether she too feels I'm intruding into a private domain. Tom stirs but his eyes stay shut, so I go on quietly, not wanting to disturb him if he's asleep. I don't want to leave yet, but neither do I wish to wake him up.

I wonder to myself what I'm doing there, but before the thought can be well-formulated Tom opens his eyes and says "Hello" to me. I repeat that his physician wants me to talk with him. Just as I'm about to say why – which I figure he probably suspects anyway – his eyes roll up and his face contorts, his limbs flail about. His mother quickly turns to him, touching his brow, his arm, trying to soothe him, his withered legs still covered by the blanket, her face marked with apprehension and concern. She looks up at me and I quickly retreat: "I'll get the nurse . . . ," I say. I wonder how my words could have brought on such agony. Soon I realize how stupid that thought is when the nurse calms me down as she sweeps by me. She says he's having another seizure. I back out with the words about returning tomorrow.

Walking into that room I noticed immediately the patient's centering presence: Tom, doing nothing but lying there, yet drawing my gaze straight away. His mother is seated, yet is postured toward Tom. Her hand confirms my gaze: *he* is the center. Everything is oriented toward him. Though lying still he highlights the fact that everything that goes on is geared toward him. His prone body is the focus of all the gauges, tubes, lines, and other gadgets such rooms display. Visitors are in the way. They have to get out of the way whenever nurses or doctors enter and move immediately to that centerpiece, the patient.

Talk, too, centers on the patient. If able himself to talk, as was Tom, it's all directed to and about him. Even the doctors' and nurses' talk swings everyone's attention to the patient. Others in the room are accessories, incidental to the scene and the action. Talk, even small talk among the others, converges on the patient: What's going on? What's wrong? How long will he be this way? What can be expected to happen next? "The day is lovely," you say, trying to lighten the room's mood, but you say it knowing the patient can barely tell what makes it lovely. So you talk about the sun, the flowers, the cool temperature – and the patient smiles, brought for a moment out of himself, his pain, uncertainty, and suffering.

Bodily gestures are muted in the patient's presence, as if being able to move about and gesticulate freely is insulting – an unwelcome reminder of things once within his repertoire but now beyond. This is especially true of Tom who cannot walk. You feel somehow uncomfortable walking in front of him. But he doesn't particularly notice, and I sense my own awkwardness in moving so slowly about. I scratch my head, wondering all the while whether he can do it, too. Of course he can, but I can't keep from going on silently to myself: with his hydrocephalus and shunt, what does he feel when he scratches his head? Has he become habituated to the knobbiness and puffiness there? As have I to my balding pate?

When he convulses, I try to unsay what I've just said, thinking, what in the world does ethics have to do with whatever is happening to him now? I try to unwalk my way back out again but am trapped by his mother's pleading glance at me. She pins me to the place I've reached but then she pulls back to look at her son writhing on the bed. I, too, am drawn to look at him, wanting somehow to undo whatever I've done (if anything). I want to help somehow, but there's nothing I can do, or should do, except get someone there who can do something: the nurse. She glides in and around me – an incidental to the scene, I move to get out of her way – and moves straight to Tom. His mother, too, withdraws her arm, moves her body back to her chair. Her look is one of concern and worry.

My thoughts are scattered and need refocusing. I retire into myself outside his room, trying to collect my thoughts, to collect myself. A theme for tomorrow's encounter is needed but what could it possibly be? Then the attending's request comes back to me and things settle back into shape. I know what must be talked about but today's scene has been rattling, unnerving. Tom's body, I think to myself, is one of those spina bifida babies all grown up; his legs, useless to him, can't be seen but I know what they must be like. His upper torso looks at once powerful and soft, well-developed yet somehow doughy. His arms are powerful, clearly having had much work to do over the years propelling the wheel chair.

Now I realize that his upper torso is outsized. There's a kind of potency concentrated in his shoulders, arms, and hands. He's become habituated to gripping and shoving. He looks his age, about 28 or so. His face is unmarked by any wrinkles. When first he speaks, there is hardly any expression played out. Everything seems rather focused in and on his mouth, his lips. Then the seizure; his face, warped with grimaces, distorted by what at first seemed raw pain – all involuntary, I knew afterwards. What must it be like to undergo that?

4. CLINICALLY REFLECTING ON THE OTHER

To walk into such a hospital room is to undergo an experience as fascinating as it is transformative. Everything is *oriented* in a quite specific and powerful way: glances, talk, gestures, equipment, attention are all placed around and directed to the patient. This is in striking contrast to the structural imbalance of the relationship – it is *asymmetrical* with power on the side of the physician. The physician, not the patient, has the knowledge and skills to help. The physician, not the patient, has access to resources (diagnostic technologies, prescription drugs, surgeons, hospitals, and all the rest). Physicians are legally authorized and socially legitimated to use their knowledge and skills for patients. In this respect, not only is the patient compromised by illness, but is also disadvantaged by the very relationship to those who profess the ability to help [12].

Nevertheless, entering such rooms, the realization dawns on you that the patient clearly dominates: Tom is the center of attention. Thinking about this, it also comes over you that these notions – being-oriented-by the patient's presence and the patient's being the center of attention and action – are complex in a way that can be delineated.

The afflicted person is compelling, that is, whether the person is ill, injured, or crippled by an accident of genetic or congenital circumstance (for shorthand, I'll merely use "ill"), he or she exercises a peculiar attraction on others. It is almost as if a kind of gravity pulls others' glances, gestures, and talk toward the ill person. A feeling comes over you that you want, even have, to do or say *something*. At the same time, to be drawn to the ill person is to feel intrusive, as if the mere act of looking at him or her were somehow to trespass on forbidden terrain. This feeling remains unless the patient indicates that it's okay. It's as if talk has the shape of an apology – a preliminary asking to excuse the trespass. And even with permission, it's hard to know just what to say, and how to say it. Why is this?

The afflicted person presents as exposed and vulnerable, that is, in a remarkable dialectical opposite to the usual sense of gravity or force-of-attraction, the ill person attracts us precisely by and through presented *difference* – which is to say, by his or her vulnerability. There is a vibrancy, an aliveness one senses in the way people usually present: the look of the eyes (welcoming, threatening), the lift of the flesh (face, hands, movements), the spoken word (pauses, emphases). This aliveness is muted in illness – even when, as with Tom, pain or seizure drive sudden, forceful gestures and bodily displays across his face and torso [11]. With grievous or terminal illness in particular, the patient's body takes on a pallor, a kind of wanness that provokes one to wonder whether the person himself is still vibrantly present within that embodying body.

Tom's vulnerability was clear and unmistakable: abed, begowned, monitored, betubed. Yet the exposure which leaves the patient asymmetrically open, readily-available for medical examination, nursing ministration, and other technical touchings, is commanding and compelling. You feel it immediately on entering the room: *Don't touch! Watch what you say! Wash your hands! Don't sneeze!* Somehow, this vulnerability is powerful. It attracts, directs anyone who approaches the patient to *be careful* – in words, touches, glances. Before the ill person we are brought out of ourselves and drawn toward him or her. What can or could be done is silently governed by *restraint*. Ancient physicians termed this *sophrosyne* and, blended with *dike*, understood these as the prime virtues of the "art" ([1], pp. 6–24). But any person feels this attraction-with-restraint when encountering the ill person: don't take advantage of the sick person *precisely because he or she is sick*. Vulnerability presages care, compels cautious attentiveness, and prompts uncommon sensitivity to hurts and harms the person is already undergoing.

The ill person evokes a moral sense. Analyzing Albert Schweitzer's "other thought in ethics," Herbert Speigelberg argues that Schweitzer was not so much formulating a principle as appealing to modern man to broaden and deepen his moral sense, to awaken "a moral sense that is usually dormant but that on special occasions can be brought to the surface" ([8], p. 232).[3] The encounter with an ill person is precisely one of those special occasions and entering a room like Tom's, you find yourself feeling that very sort of "awakening."

These feelings, moreover, are specifically *focused*: attention is directed-to the sick person ("How do you feel?"), oriented-toward the patient's particular circumstances ("What's wrong with you?"), and *aimed-at* his or her possible futures ("When will you get better?"). This affective ground is an almost visceral alertness – a bodily tug to be mindful of this person within the actual context of his compelling vulnerability. As directed-to, oriented-toward, and aimed-at, these feelings propel the self beyond itself. They are in this sense an elemental *ec-stasis*; to be self is to be-beyond, always-already with the other. In this, "subjectivity" is from the outset of life "co-subjectivity." Subjectivity (selfhood) is an accomplishment; it is a prize won through complex developmental experiences.

These feelings are, as it were, "come upon"; sensing them, you seem more to find yourself in their grip than to plot or plan them out ahead of time. It's not even proper to say, "I feel . . . ," as it is rather a matter of "it is felt in me." To the extent that one thinks about it, it's not so much that "I think" but rather that "it is thought in me." These feelings *happen to* the patient's visitor. But whether you permit them space and place is always another question. Drawn by these feelings to the other-who-is-ill, this being-oriented-to the other (*Du-Einstellung*, in Alfred Schutz's term [9]) is a kind of *invenire*, whose core meaning is "to come-upon" or "to come-into" ([10], pp. 21–22, 33, 195).

The patient presents as a kind of question: Do you care? Does my being here and out of things, if only for a while, make any difference? Do I matter? As the patient is a kind of question, so does the happening of those orienting feelings occur as a felt need to respond: to be *responsive to* the ill person, here and now, and to be *responsible for* the response (whether word, touch, gesture, glance, whatever). I am drawn *by and to* the other *as* myself. The ill person attracts as a vulnerable question, and compels response: this is the fuller sense of the *ec-stasis* that embodies the visitor's being-oriented-toward the other.

5. MEDITATING ONTOLOGICALLY ABOUT THE CLINICALLY PRESENTED OTHER

This *ec-static* moment is deeply embedded, embodied, inscribed inwardly as what it fundamentally means to be self.

Embodiment: that "intimate union" that so frustrated Descartes and so many others after him. So "intimate" is self bound to its embodying body that there is the constant temptation to say, "I *am* my body": hit "my" body and you hit "me." Here is precisely the source of "belonging," from which all other senses are derived ([11], pp. 47–66). Yet, however intimate and profound is the relation between the self and the self's body, it is equally true that the body is experienced as strange and alien.

I *am* my body; but in another sense I am *not* my body – or not simply that. This otherness is so profound that we inevitably feel forced to qualify the "am:" it is not identity, equality, or inclusion. It is "mine," but this means that self is in a way distanced from its body, for otherwise there would be no sense to "belonging," to experiencing it (and other things) as "mine." So close is the union that self's experience of the own-body can be psychologically surprising (its happy obedience which the self notices for the first time), even shattering (its hateful refusal to obey my wishes to do something: walk, jog, or whatever). So intimate is it that self has moments in which it genuinely feels "at home" with it. Yet, so other is it that there are times when self treats the own-body as a mere thing that is *other* (obsessively stuffing it with food or otherwise mistreating it; or as when it is encountered as "having a life of its own" to which self must willy-nilly attend: like it or not, "my" hair grows and must be trimmed for certain purposes, "my" hands cleaned, "my" bowels moved, "my" cold cured, and so on) ([11], pp. 35–55).

Embodiment is an essentially *expressive* phenomenon. It is that whereby feelings, desires, strivings, etc., are enacted (albeit in culturally and historically different manners). As such, embodiment is *valorized*. After all, what happens to *it* happens to *self*. As that whereby the person "rules and governs," self is at the same time subject to its conditions. What happens to it thus *matters* to the self whose body it is: the embodying organism lies at the root of the moral sense of inviolability of self, of personhood – of the "privacy," "integrity," "respect," and "confidentiality" that play such profound roles in bioethics and clinical ethics. Nor does the fact that people can and do dissemble and deceive themselves and others belie the expressivity of the body. Indeed, these are themselves expressive phenomena, however difficult it may be to discover and then to interpret them.

This value character of the embodying organism also helps elucidate more fully why the continuing discussions of many bioethical issues – pregnancy, prenatal diagnosis, abortion, psychosurgery, withdrawal of life-supports, euthanasia, etc. – are so highly charged and deeply personal. It is, was, precisely what made the encounter with Tom so puzzling and difficult. On the other hand, the profound moral feelings evoked by certain medical practices (surgery, chemotherapy, dialysis) and much experimentation (for instance, the human genome project) are understandable, as they are in effect ways of intervening or intruding into that most intimate and integral of spheres: the embodied person. Self is embodied, enacts itself through that specific animate organism which is "its own" and is thus expressive of self. Bodily schemata, attitudes, movements, actions, and perceptual abilities are all value-modalities by which self enacts and expresses its character, personality, habits, goals – in short, by which the self is alive as such.

As alive, the embodied self is intimately bound to death; when born we are old enough to die. At the same time there is something else about human birth. Unless nurtured by others – most obviously, parents – the baby most surely *will* die. Unlike other animals and most primates, human beings are born too early ([7], pp. 356–357). We do not come ready-equipped at birth with the repertoire of instincts and abilities necessary to make it on our own. Sociality is fundamental to human life, far more thoroughly than often meant. The typical opposition, "nature *versus* nurture," is never more than a half-truth.[4] Whatever may be our specific biologic endowment, "being" is "becoming" (as Gabriel Marcel said, to be human is "*être-en-route*," "being-on-the-way" [6]), and becoming is necessarily a matter of being enabled-to-be by and through a myriad of nurturing actions by others. The other is inscribed and dwells within us, enabling us to be what we are. To become a self and eventually a person requires multiple and multiply complex interrelationships with other, already-developed persons ([11], pp. 144–181).

These relationships are *reflexive*. Kierkegaard was right: to be self is to be reflexive. Indeed, self is always a reflexive expression divested of the pronominative it modifies. Self is thus related to another. More accurately, it is that in the relation that the relation relates itself to its own self ([4], pp. 146–147).[5] So understood, however, it is still not completely grasped, for "*that* in the relation *that* the relation relates itself to its own self*" is, as he put it, constituted either by itself or by another. As the first possibility is incoherent – such a *causa sui* cannot exist, cannot bring itself into, nor maintain itself in, being – the relation that relates itself within that relation to itself must have been constituted by another. Kierkegaard moves to a "Power" that, "as it were," lets this peculiar inwardly-outward reflexive relatedness "go out of Its hand" and thereby lets it be on its own. Lest that capital "P" mislead, however, one can in more mundane terms remove the "P" – and we are then confronted with the phenomena of *parenting and birth*: mother lets baby go out of its womb (better: baby finds itself biochemically ready to exit and, like it or not, there it is, *worlded*). Baby is enabled-to-be whatever it may be already within the womb, where there are already the beginnings of those mutual relationships.

Each human being is a reflexivity: self is self-reflexive however minimally this may be present at any particular stage. Its being/becoming is "staged"; it is temporally phased in its being-what-it-is.[6] In different terms: *within* the relation, mother is reflexively related to and as herself by being related to baby; baby is reflexively related to and as itself by being related to mother; and both are related to and as each other by being self-and-other-related within the relationship itself. There are thus two self-related reflexivities related, within their relation, to one another. They are bound to and as each other inwardly – that is, mutually.

By enabling the baby to be-*en-route*, mother is at the same time enabled to be-herself (i.e., she is mother by and as mothering or nurturing baby) – and conversely. Both are profoundly marked by these reflexively complex, temporally on-going, and nurturing relationships. To be self is at root to be enabled-to-be-self-aware, to whatever extent and in whatever way it may be. If that is so, and if it is true as well that this self-aware infant cannot continue to be without the enabling, nurturing mutuality relationships of mother with it, then its being as self-aware is constituted through those relationships with mother and correspondingly, mother is constituted as mother through her multiply, mutually-relating relationships with baby. Each, in Kierkegaard's idiom, is a "*that* in the relation *that* the relation relates itself to itself": *in* its very relatedness to itself it *is* self-related to the other. The already developed other is inwardly present within the developing self already from birth (and doubtless from some time prior to its being worlded). To use a simple example: Mark and I are friends; each relates to, experiences, and values the other (I like to be with him); each relates to, experiences, and values himself within the relationship to the other (I really feel good being around him); and, each relates reflexively to, experiences, and values the relationship itself (We have a good friendship).

The *ec-stasis* is a "being oneself by and as being with others." The ecstatic being is one whose inwardness and awareness of self *is* enabled by the other. To be human is thus to be a reflexive inwardness turned (by the mutual relationship with the other) reflexively outward from the outset of life. The immediate nurturing other (parent/baby) is already within the self (baby/parent) as that whereby the baby is at all able to be and become. Subjectivity *is* intersubjectivity (*esse* is *co-esse*).

6. BEING-WITH THE ILL PERSON: 'COMING-INTO' THE MORAL ORDER

Encountering the afflicted other person, one comes into or "happens" on the moral order. The ground of ethics is the reflexive relatedness to and with the other person – perhaps most poignantly presented when we come upon the other-as-stranger, and even more so when the stranger is ill.

Illness itself is a strangeness. Even while each of us is familiar with being sick or injured, to come upon the sick person is to find oneself ineluctably facing an unknown and a challenge, a dare, an appeal. With the terminally ill, the scene is set even more dramatically: Who, what, are you now? What can I do? What should I do? How should I act? What should I say? Standing bodily before a dying person whom you do not know, what *can* you do? Perhaps all that can and even should be done is simply to be there: mutually interrelated by means of touch, feel, word, look. Each of these simple gestures is precisely what it is: an affirmation of the person as *worthy*. You make a difference; you *matter*.

There is more to this, for the sick person's challenge or appeal is for a response. Most basically the appeal is to *be with* the other as an affirming presence of the person's continued worth despite illness – which is not only alienating and debilitating, but at the same time is experienced by the ill person as demoralizing. Being with the ill person is, or can be, therefore, a remoralizing. As many studies have shown, patients want most of all to know what's wrong, and, equally, whether those who take care *of* them also care *for* them [2] – do I still matter?

Being-with others is so pervasive in our daily lives that we rarely give it a second thought. The other side of this being-with-others is equally taken for granted without thinking: "put yourself in my shoes!" Each of us knows this well and often takes up that perspective, as even slight reflection shows. The patient especially presents this challenge and in so doing is not asking you to *be* the patient, even were that coherent. Nor is he asking you to think and feel what things would be like if you were sick like he is sick – though such imaginative acts can be instructive. Nor is the appeal for you to try, even were this possible, to be that person literally. The fact is that you are not that patient, he is not you. None of us truly knows anyway what we would do, think, or feel, were we in that patient's circumstances literally.

To "put yourself in my shoes" is, quite simply, to do what is most natural for each of us, if what I've suggested above is at all on the right track. Each of us is what he or she is solely within the multiple, reflexive relationships with others, as each of us has been enabled-to-be what and how and who we individually are only thereby. The presentation of the ill person is an invitation to the visitor to be precisely what and how and who the visitor is, and from and with that to think and feel what this patient faces. Each of us is always and essentially with and by means of others, including this unique other person who is ill. As reflexively, mutually related with each other (as in the example of friends above), the appeal is a challenge for each of us to recognize that even in the relation to Tom there is an intimacy, a knowing that can, if allowed, affirm the what and how and who of Tom and thereby yourself. This is the core of that "special occasion" Schweitzer identified through which a moral sense that is usually dormant can be brought to the surface – the awakening of the ec-stasis which is our being.

Another word for this shift to the other's perspective is *trust* (even if, as in most clinical situations, all that is possible is a kind of temporary trust). Tom's words and demeanor is an appeal for us to feel-with him; it invites (and thus challenges) us to *affiliate* with him and this act, I am led to think, is at the core of the moral order. In more humble terms, this affiliative feeling or felt mutuality is the dyad care-and-trust: *compassion*. To take care of a patient is, if only minimally, to care for the patient and this invokes affiliating, feeling with the patient and may assume quite specific forms – such as *respect* with its correlative enablement of dignity or integrity which is the allowing/enabling of the other to be precisely what and how and who he is: Tom, with all his blemishes and failings, his tactlessness and, yet, dignity. He is *one who matters*; his life is *worthwhile*. These specific forms of affiliation, moreover, themselves take on highly individual, embodied stances: talking and listening, touching and being touched, looking and being looked at, thereby affirming the other's humanity and worth – in turn receiving affirmation of oneself and one's own worth. The "matter" of embodiment, if you will, is the "stuff" of value, not the merely measurable stuff of physical extension.

To meet the other as ill is to meet a challenge to embody and enact the moral order, in the most concrete ways of the flesh: it is to enter, to "come-upon," the moral order and, encountering the other, to recognize and perhaps affirm the other's appeal by means of responses tuned into the specific other person *vividly*, being bodily with the other who is bodily with me, thereby constituting the core sense of community. *We, thou and I, matter to one another.*

7. RETURN

The last time I visited Tom he was on the dialysis machine. He shared with me that his boss had told him he could have his old job back. He was very upbeat, joshing about the machine, joking with his nurse, and offering again to come to one of my classes and talk about himself. I didn't have the heart then to ask him why he had earlier refused to be precisely where he was now. Nor do I have whatever it would take to ask him now. Maybe later, when he has (or I have) had time to think about things, when time can work its subtle alchemistry on our memories and deepest feelings about ourselves; then, perhaps, I can ask. In the meantime:

I suppose there might have been a time
  – a time of apples and children
  of women and glances
  of rain, snow and wild winds –
When things now usual and plain
Were only themselves, shining;
When in innocence we could let them be,
As be they might, driven in on our eyes
Like eagles, or in silence shared awhile
With no need of words. . . .
Beneath a sheath of trees . . .
A time of talking and whispering together
Like faintly shifting leaves.
Or singing slipping through the quiet air
Like birds' wings,
With minds like picnics spread quietly around
A mound of soft grass.

If such times of knowing (being) are
What it's all about, then what is time?
Quick paradox of inwardness: yearning
On the further edge of living, for what was,
And ought to be, or to have been. . . .

To whom offer these sly celebrations of memory?
On whom depend for understanding where we stand,
Beneath the bows of trees, lying, on our backs,
Thinking of home and apples, children and glances,
Listening for voices, like birds sleeping in leaves,
To tell us, now, in this needful time, the time we are,
Who we are, or why?

    I'm not sure why Tom remains so firmly in my mind. Neither deft nor subtle, he was in truth – and I'm sure is still – unvarnished. It's not so much his refusal that stands out, as the harsh accident of his birth, the turmoil of his life. That utterly fortuitous and sly genetic accident that destined this gentle, artless man to face, unasked, a life marked by curious nostalgia for what he could only guess at, yearning always to be *normal*.

Vanderbilt University Medical Center
Nashville, Tennessee
U.S.A.

NOTES

[1] This narrative is adapted from one of those in my *Troubled Voices* [13].
[2] I had come to know Donald ("Dax") Cowart some years ago, but this was years after his accident. Dax had made every effort to refuse treatment for severe burn injuries that left him seriously handicapped. He continued to be treated against his will. His story became well known in ethics and medical circles and he had been my guest a number of years ago during a conference held in Nashville. There are two records of his case. The first, an underground classic in the 1970s, is a video made by Robert White, entitled "Please Let Me Die," while Don was still hospitalized in Seely Hospital in Galveston, Texas, in 1974. The second, "Dax's Case," is a film made in the early 1980s, available from Concern For Dying, Inc. and produced by The Texas Committee for the Humanities, under the direction of L. D. Kliever. A fine volume of essays was subsequently edited by Kliever [5].
[3] That "other thought in ethics" is, Spiegelberg points out, that one's "good fortune" (health, talent, ability, family, etc.) obligates, constitutes a "special responsibility": to give something in return for that good fortune ([8], p. 232).
[4] One could say either "nature nurtured" or "nurture natured." Whatever cultural or familial variations there may be, these must still occur within limits established by one's biological endowment; conversely, one's biological endowment, while it surely includes detailed genetic instructions, for instance, is nevertheless articulated within a specific cultural and familial setting that itself promotes, channels, and mutes certain bodily expressions, actions, etc.
[5] "[T]here can be two forms of despair properly so called. If the human self had constituted itself, there could be a question only of one form, that of not willing to be one's own self, of willing to get rid of oneself, but there would be no question of despairingly willing to be oneself. . . . [Hence] the self cannot of itself attain and remain in equilibrium and rest by itself, but only by relating itself to that Power which constituted the whole relation" ([4], p. 147). "Despair" itself, as a "disrelation in the relation that relates itself to itself," accordingly, arises by virtue of the "the relation wherein the synthesis relates itself to itself" ([4], p. 149).
[6] In a wonderfully lucid essay on Marcel, William Ernest Hocking points out that the best way to render Marcel's notion of ecstatic "being" is the New England active usage: "I be. . . ." This strikes me as right on target, at least for what I'm driving at in the text. I would add, however, that this "I be . . ." is essentially "I be-with-thou"; the "I" is staged into "being" by and through the Other, as Other "is" by and through the staging "I" ([3], pp. 441–445).

BIBLIOGRAPHY

1. Edelstein, L.: 1967, *Ancient Medicine, Selected Papers of Ludwig Edelstein*, Owsei Temkin, and L. C. Temkin (eds.), The Johns Hopkins University Press, Baltimore, MD.
2. Hardy, R. C.: 1978, *Sick: How People Feel About Being Sick and What They Think of Those Who Care For Them*, Teach 'Em, Inc., Chicago.
3. Hocking, W. E.: 1952, 'Marcel and the Ground Issues of Metaphysics', *Philosophy and Phenomenological Research* XIV #4, 439–469.
4. Kierkegaard, S.: 1954, *The Sickness Unto Death*, Princeton University Press, Princeton, NJ.
5. Kliever, L. D.: 1989, *Dax's Case: Essays in Medical Ethics and Human Meaning*, Southern Methodist University Press, Dallas, TX.
6. Marcel, G.: 1951, *Le Mystére de l'être*, Tome 1, Éditions Montaigne, Aubier, Paris.
7. Portmann, A.: 1953, 'Biology and the phenomenon of the spiritual', in J. Campbell (ed.), *Spirit and Nature: Papers from the Eranos Yearbooks*, vol. I, Bollingen Series XXX, Princeton University Press, Princeton, NJ, pp. 342–370.
8. Spiegelberg, H.: 1975, 'Good fortune obligates: Albert Schweitzer's second ethical principle', *Ethics* 85: 227–234.

9. Schutz, A., and Luckmann, T.: 1973, *Structures of the Life-World*, vol. I, Northwestern University Press, Evanston, IL.
10. Wolff, K.: 1976, *Surrender and Catch*, D. Reidel Publishing Company, Dordrecht, the Netherlands.
11. Zaner, R. M.: 1981, *The Context of Self*, Ohio University Press, Athens, Ohio.
12. Zaner, R. M.: 1988, *Ethics and the Clinical Encounter*, Prentice-Hall Inc., Englewood Cliffs, NJ.
13. Zaner, R. M.: 1993, *Troubled Voices: Stories of Ethics and Illness*, The Pilgrim Press, Cleveland, OH.

CULTURAL DIVERSITY AND THE CLINICAL ENCOUNTER: INTERCULTURAL DIALOGUE IN MULTI-ETHNIC PATIENT CARE

Of all of the images in modern medical ethics, none has been as powerful or has stimulated as much creative analysis as that of the doctor and patient as *strangers*. Long-term relationships between doctors and patients are increasingly rare in American health care, and physicians' ancient claim to benevolent paternalism was one of the first casualties of modern medical ethics. Contemporary models of the therapeutic relationship, most notably the covenant and the contract, start from the premise that the doctor cannot presume to share the patient's beliefs or goals, and that both parties must come to know each other during their limited time together through meaningful communication and clinical negotiation.

Although the theoretical frameworks of covenant and contract provide valuable critiques of traditional medical ethics, experience demonstrates that neither takes the image of the stranger seriously enough for meaningful application in a multicultural clinical setting. Culturally based presuppositions and expectations about illness, appropriate treatment, and the sick role are much more diverse than their vision of pluralism acknowledges, and their proponents' admonitions for doctors to respect patients' self-determination raise more complicated questions than these theories typically accommodate.

The importance – and the difficulty – of developing an intercultural approach to ethical doctor-patient interaction has recently gained prominence in the ethics of international research [3, 5, 17, 34]. While this issue may be new to many ethicists, there is a massive literature in medical anthropology and sociology describing the complexity of nonwestern medical systems and relationships, and the conflicts that can result both here and abroad when patients and practitioners from disparate cultures come together. As the experience of clinical ethicists becomes more influential in ethical theory, the value of a social-scientific approach is more evident: understanding the practical sociocultural context of values and behavior is essential to achieving ethical ideals.

North American ethicists' standard view of the ideal doctor-patient relationship in a pluralistic society – a negotiated alliance of mutually respectful, autonomous individuals who act out of enlightened, rational, self-interest toward a common goal – has itself been tempered by clinical experience. A number of clinical ethicists have observed that real people, however knowledgeable and self-confident while healthy, typically become powerless and uncertain as patients, and must be encouraged by practitioners to assume authority. In every therapeutic relationship there is an element of unavoidable vulnerability and trust, where the patient is at a disadvantage before caregivers, even as he or she wants to be self-determining [71]. This tension

G.P. McKenny and J.R. Sande (eds.), Theological Analyses of the Clinical Encounter, 203–223.
© 1994 *Kluwer Academic Publishers. Printed in the Netherlands.*

between trust and self-determination increases the more the doctor and patient are strangers to each other.

The highly touted answer to the challenge of pluralism is good communication: typically the physician is enjoined to provide the patient with appropriate information and reassurance, elicit the patient's values, opinions and questions, and share decision making through open, honest negotiation. While good communication is essential to any meaningful therapeutic encounter, the nature of communication is also culturally determined. The standard western view of good doctor-patient interaction is also dependent on autonomy-based conceptions of communication that assume not only that both doctor and patient are conscious of their own cultural presuppositions, but that they are both able and willing to articulate them in a way that is meaningful and persuasive to the other. Again, clinical experience in a multicultural setting demonstrates that such assumptions are often incorrect.

Persons from nonwestern cultures, particularly those from nonindustrialized, hierarchical societies, are not well served by a view that defines the appropriate patient role as that of a self-interested, self-determining individual who must actively assert his or her goals and opinions. For many such patients, individuality and autonomy are not ethical ideals, perhaps not even meaningful concepts. The conviction of medical ethicists that health care providers can and should work to empower their patients by encouraging them to take an active role seems not only impractical in such an intercultural setting, but ultimately destructive of the very values that it is intended to promote.

This essay examines the issue of divergent cultural beliefs and values in the therapeutic encounter, and the questions that both caregivers and ethicists face in determining how best to respect patients from nonwestern cultures while remaining true to western ethical ideals. Although medicine has often sought to distance itself from images of religious belief and practice, both medical practitioners and ethicists can find a valuable model for intercultural patient care in interreligious dialogue. Such dialogue in medical ethics and clinical medicine can offer new approaches to understanding individual patients' particular needs and new challenges to our conception of ethical relationships in a pluralistic world.

CULTURE AND MEDICINE

Culture is a dimension of human life in which everyone participates, but which few can define. For the purposes of this essay, culture refers to a system of shared beliefs, values, and behaviors that are a product of group experience, interpreted in light of beliefs about the purpose and meaning of life. Race, language, nationality, geographic origin, and religious background are important dimensions of culture, but none is essential to cultural identity [61]. Both groups and individuals have significant investment in claiming membership in a particular culture. However, the expression of culture may vary markedly among subgroups, among individuals within subgroups, and over time.[1]

Typically, fundamental elements of culture, such as gender and age roles, dietary practices, and means of self-expression, are so integral to daily life that they are frequently taken for granted as human nature. Much of the meaning that is attributed to behaviors and beliefs is symbolic rather than literal, and cannot be easily understood or articulated apart from lived experience [21, 30, 61]. Such factors become apparent often only when they come into contact, and conflict, with the seemingly inappropriate behavior and beliefs of other cultures. Even then, the resulting culture shock may be overwhelming, making analysis of one's own presuppositions difficult [57, 66].

One of the United States' most important social myths is that of the melting pot, where diverse cultural groups blend in a society that is at some point unified in goals and values by the American experience. The image of the melting pot has been strained in recent years. As millions of immigrants have come to the United States in the last two decades, often from lands unknown to many Americans, the American experience itself has changed; by the turn of the century, the "ethnic minorities" will be the majority of the American population. While acceptance of diversity is generally cited as an important element of American culture, widespread understanding of the many American subcultures remains elusive.

The high rate of immigration has had significant, albeit subtle, effects on the U.S. health care system, creating new occasions for intercultural contact and conflict.[2] Health professionals, especially those who have studied in North America or Europe, have been some of the first immigrants from many countries. Many foreign medical graduates go to small towns and rural areas that have difficulty attracting physicians, and whose residents may know little about other cultures [26, 39]. Even in large cities, however, many patients may be unfamiliar with a foreign doctor's heritage. The nursing shortage has also increased the ethnic diversity of professional caregivers. The number of immigrant nurses providing bedside patient care has risen sharply, as hospitals and government health authorities have actively recruited nurses from other nations to fill empty positions [19].

To some extent, years of medical education and socialization create a professional culture of medicine that diminishes the effects of caregivers' particular heritage on the treatment that they provide [31]. However, even among western, industrialized nations, what is taught as medical science varies significantly according to cultural values essential to national identity [54]. More important for the consideration of patient-practitioner relationships, immigrant caregivers are likely to retain their original cultural presuppositions about social roles and behavior, despite their westernized practice of scientific medicine [30].

More visibly, the high rate of immigration has increased the number of ethnic minority patients, many of whom are poor and must be treated in already-strained public facilities. Depending on their reasons for emigrating, they may suffer from serious physical and/or psychological conditions that require long-term medical attention [63].[3] Moreover, they may need treatment for conditions that are rarely if ever seen among nonimmigrants, such as rickets or leprosy, and may occasionally carry diseases that pose a public health risk, such as tuberculosis or cholera.

When an immigrant or ethnic minority community has no identified physician of its own, patients may be unable to identify the medical practitioner who is best suited to treat them [41]. Additionally, their access to health care is typically restricted by a lack of insurance, a poor understanding of the medical system generally, and often a limited command of English [1] (E. Heitman, unpubl. PhD dissertation, Rice University, Houston, TX, 1988). Typically such patients rely on home remedies, traditional healers, and pharmacists, even for serious problems. If their conditions become unbearable, many seek treatment in hospital emergency rooms. However, emergency room staff are frequently too busy to give personalized care, and are often resentful of the additional services that such patients may require; these factors reinforce patients' reluctance to seek a physician's help in the future [1].

The scientific medicine of North America and western Europe has its own version of the myth of the melting pot. Western medicine is widely perceived by its practitioners to be a universally objective endeavor: the effectiveness of X-rays or diuretics or hip-replacement surgery is dependent solely on biological factors, irrespective of cultural variables that affect the patient or practitioner. Thus "good medicine" is defined as the objective application of western scientific principles, which should provide uniformly appropriate diagnosis and beneficial treatment for all patients.

In the development of contemporary theory on therapeutic relationships, ethicists have typically decried physicians' objectification of their patients and overreliance on technological means of knowing. However, ethicists have accepted uncritically western medicine's belief in objective knowledge. For many, "ethical medical care" in an intercultural setting depends on an objective process of interaction where doctors and patients negotiate appropriate relationships and treatment; the process is thought to be universally applicable because it defines the content of the relationship as the situation demands [12].

However, this theory's applicability in intercultural medical care is limited by its culturally determined faith in objectivity: it assumes that doctors and patients can abstract health care from the worlds in which they live, and define objectively the elements of the encounter to be negotiated. To the contrary, because medical beliefs are so deeply embedded in culture, the meanings that each party attributes to health and illness, diagnosis and treatment, and the roles of patients, caregivers, and communities may not be amenable to such objectification and analysis.

INTERCULTURAL COMMUNICATION IN THE CLINICAL SETTING

The honest, rational communication that is posited as the means of defining the terms of relationship is also understood by many as a largely objective enterprise. However, even where patients and practitioners have a common background, their communication is seldom direct, and the information that they exchange in direct conversation seldom reveals the more fundamental assumptions that each makes about their purpose and goals [9, 20, 37]. More often, essential beliefs, hopes, and fears remain unspoken, even if recognized. The total message that each receives in the encounter is both much more and much less than the words that they use. The possibility for miscommunication grows precipitously when doctor and patient have culturally different styles of communication.

Direct, honest communication is itself a practice almost exclusive to western societies [10, 24, 27, 57]. Asians typically understand direct questions as intrusive, and may be reluctant to reveal personal information that they regard as private. Asians, Arabs, Indians, and Africans are often unwilling to disagree openly or to say "no" to another's request, believing that such a response is a rude breach of social harmony; tentative agreement is especially important for preserving relations with persons in authority or from whom one seeks assistance.

Whereas westerners believe that speaking clearly and maintaining eye contact is a sign of honesty and trustworthiness, Hispanics and Asians typically interpret forthrightness as hostility; they speak softly and avoid eye contact out of respect. Beliefs about the appropriate expression of emotion, including responses to pain, vary greatly: most northern European and Asian cultures value stoicism and reserve, where southern European, Hispanic, Middle Eastern, Indian, and African cultures typically encourage expression of strong feelings [27, 32].

To persons from cultures that have rigidly defined gender and age roles, westerners' attempts to treat all patients equally may appear to be anything but respectful. Female patients may be shocked or embarrassed by comments or questions that are addressed to them directly by men; male patients may respond poorly to women in positions of authority. Westerners' protective or dismissive treatment of the elderly may appear patronizing or disrespectful to persons from societies where the elderly are honored for their wisdom and experience and are centrally involved in important family decisions.

Simple body language and gestures that practitioners may use in an effort to make the patient comfortable may have the opposite effect [7, 10, 24, 33, 57, 62]. A doctor's warm smile and cheerful demeanor may be interpreted as inappropriate frivolousness by eastern European patients, who typically equate a caregiver's serious approach with taking their conditions seriously. Patients who wish to maintain a respectful distance from their caregivers may become uneasy in a small examination room, while Arab patients, who find "sharing breath" essential to good communication, may find western physicians to be distant and uncaring because they do not come close enough.

Extending a handshake to patients and their families upon meeting is essential etiquette in western democratic societies, but an offered hand may be intimidating to patients from hierarchical cultures. A casual wink, generally considered friendly in the United States, is widely interpreted as lewd and insulting elsewhere. Gesturing with certain fingers or a raised palm, or even simply pointing to parts of the body, may communicate grossly insulting messages to patients who do not share the caregiver's culturally defined body language.

Clearly, good communication is exceptionally hard to achieve when the patient and practitioner do not speak the same language [7, 10] (Heitman, unpubl.). Beyond the obvious difficulties that a language barrier creates, many patients are embarrassed or ashamed that they do not speak English, and may go to some lengths to avoid conversation. Such diffident patients may conceal their poor comprehension with smiles, nods, and the appearance of satisfaction. Moreover, the stress of the unfamiliar clinical environment or of illness itself may cause some patients to revert to their native language, despite their ability to speak English well in other settings.

Many professionals develop a repertoire of simple questions and commands in the foreign language(s) that they encounter regularly, which they use in all but the most challenging situations. Despite their momentary practical benefit, the limited success of such interactions can be deceiving, and caregivers may mistakenly believe that they have adequate information to diagnose and treat their patient's condition after exchanging only a few words. Others, frustrated by the incompleteness of their communication, may discount the seriousness of non-English speaking patients' complaints, or respond to them with hostility [28].

Except in facilities that serve an identified population of non-English speaking patients, caregivers can rarely be expected to speak such a patient's language well. In culturally diverse cities, it would be virtually impossible for any practitioner to master all of the languages that patients might speak. Whenever the physician is not sufficiently fluent in the patient's language to take a meaningful history, explain the patient's condition and recommended treatment, and answer the patient's questions, translation is essential to the encounter.

Few institutions employ professional interpreters; often, the nearest person who speaks the patient's language is drafted for the task, whether it be a family member or friend, another patient, a bilingual caregiver, or a random employee who looks to be of the patient's culture. This practice can result in severely impaired communication. The translator's identity, including gender, age, and social status, is vital to the quality of the interaction, as he or she must often serve as a "culture broker," explaining concepts as well as translating words [10, 38] (Heitman, unpubl.).

Unless the translator is a professional or otherwise certified interpreter, it is almost impossible for a practitioner to assess the translator's proficiency or knowledge of medical concepts, the accuracy of the translation, or the patient's understanding of the exchange. Patients may conceal personal information essential to diagnosis and treatment when the translator is someone whom they know, or may lie to meet that individual's expectations. The translator may also filter information from the patient to the doctor, or respond from personal knowledge and beliefs rather than translate questions and their answers.

In facilities that treat large numbers of non-English speaking patients, caregivers may expect to be unable to communicate with most ethnic minority patients. Typically the result is not a greater effort to reach these patients, but frustration, resignation, and an unwillingness to distinguish among their various levels of understanding, all ultimately leading to poor communication. Practitioners may also assume that patients with poor grammar or pronunciation in English do not understand well either, or more harmful still, that they are not intelligent enough to comprehend medical concepts.

Although fluency in a patient's language can help practitioners humanize their care, the clinician's skills are incomplete without the cultural framework to understand the patient's views and actions. Most practitioners are sufficiently unaware of nonwestern systems of health beliefs and practices that, even when they recognize a patient's behavior to be culturally determined, they cannot put it into a meaningful context or apply that knowledge in future situations [52, 57, 62]. The most essential, and most varied, culturally based beliefs relate to the one area in which medical science is least likely to expect them: the diagnosis and treatment of disease.

DEFINING ILLNESS AND THERAPY IN THE INTERCULTURAL ENCOUNTER

Worldwide, the description, diagnosis, and etiology of human ailments vary greatly. A vivid example of this diversity is the World Health Organization's (WHO) *International Classification of Diseases* or ICD-9 [71], a catalogue of the maladies that affect humanity, as described to WHO by the medical experts of the world's cultures. Western medicine acknowledges only half of its entries to be "real" diseases [35]. The others, many of which are known as "culture-bound syndromes," typically reflect nonwestern worldviews that define human health in terms of systems of balance, social harmony, and spiritual purity. Some well-known culture-bound syndromes include "evil eye," "amok," and "voodoo death." While some of these syndromes may be recognizable, treatable entities in the western system, their scientific names, causes, and treatment are quite different from those of the cultures in which the conditions are prevalent.

Despite the almost universal presence of "disease," people typically seek medical care only when they perceive their health to have deviated from socioculturally defined norms, interpreted through their personal experience with illness [3, 50, 73]. The first step in establishing a therapeutic relationship is the patient's definition of his or her condition, its probable causes, and the most appropriate form of treatment. The physician's ability to anticipate and appreciate the patient's expectations of diagnosis and treatment, and then meet them satisfactorily, is essential to the success of the relationship and ultimately of treatment [18, 24].

There are many obstacles to meaningful communication about diagnosis when doctor and patient come from different cultural systems. Despite the admonitions of ethicists to establish common ground with patients and the currency of various biopsychosocial models of health and illness [14, 23, 65], many western clinicians still ask little about their patients' worlds or beliefs about health, and few seek patients' own opinions about their presenting complaints [9, 59]. Doctors are likely to shape their questions to obtain information that applies to a scientific model, and interpret the patient's personal remarks through their own cultural filters.

When patients' views are culturally unintelligible to their doctors, physicians typically base their treatment on the conviction that scientific medicine is universally applicable; often they assume that they can resolve the problem by treating the objectively verifiable, physical disorder. Yet, patients seek treatment that is consistent with the worldview that defines their condition. Patients are most likely to be "noncompliant" when the recommended treatment does not fit their expectations or personal sense of meaning, and are likely to reject an unfamiliar diagnosis and related intervention [13, 24, 62]. The doctor's failure to recognize the patient's view of the problem, or inability to understand its internal logic, jeopardizes not only the patient's compliance with treatment, but also the patient's trust in the doctor's qualifications. Such doubt threatens the success of the treatment in question and any future relationship with other physicians.

Thus, a mother who describes the onset of her infant's "evil eye" will likely be dismissed as uneducated and superstitious by a doctor who seeks a primarily biological cause for the child's fever, restlessness, and prolonged crying.[4] The mother will not be content with treatment for an ear infection, and will likely not follow the prescription; rather, she will look for a way to remove the curse that created the physical symptoms. Her confidence in the physician may fade because of the doctor's failure to recognize evil eye, and apparent inability to treat such a common, and threatening, childhood condition.

In some situations, the conflict may lie not in the definition of the patient's ailment, but in cultural definitions of what sorts of deviance constitute disease. Western doctors may be perplexed by a North African man's complaint of amenorrhea unless they know that his entire community suffers from the chronic urinary tract bleeding of schistosomiasis. Conversely, an Brazilian emergency room patient who believes that grand mal seizures are a sign of her calling as a spiritist medium is unlikely to accept a neurologist's diagnosis of epilepsy or a prescription for anticonvulsants.

Whereas scientific medicine defines the causal mechanisms of disease in biological terms, most lay people and nonwestern medical systems interpret illness in moral terms as well [2, 32, 38, 62]. In all of the world's traditional medical systems, illness has been linked to personal moral transgression, whether as sin, impurity, karma, or disruption of social harmony [32]. When the biological origins of a disease are also associated with morally questionable behavior – as in the case of alcoholic cirrhosis of the liver, it can be particularly difficult to separate conclusions about biological causality from those about moral causality. Many patients are reluctant to disclose important aspects of their experience with illness out of shame and fear of others' moral judgment [2, 45]. Unless the physician is aware of and anticipates patients' culturally determined moral presuppositions, the doctor's questions may paradoxically lead some patients to withhold information or lie outright in an effort to avoid being judged.

Shame and humiliation are especially associated with parts of the body that are often the focus of medical attention, and the varied cultural symbolism of the body can be the source of considerable misunderstanding between doctors and patients. Although the bowels and genitalia are almost universally considered private, there are strong cultural prohibitions in many societies against discussing sexual or excretory organs and functions; among many Indians and Pakistanis, serious problems with sexual or excretory functions would not be mentioned even between spouses [57]. Many Asian, Hispanic, Indian, and Middle Eastern women find gynecological examination to be a highly shameful personal exposure [58, 64].

Other parts of the body have cultural and religious importance as well [16, 33, 50, 62, 72]. For many Buddhists, the soul rests at the top of the head; touching another's head unnecessarily or without respect is grossly insulting and invasive. Among Moslems and many Buddhists the foot is considered to be unclean, its display to others is insulting, and its examination by others humiliating; however, for some Africans and Native Americans, the feet are the conduit of life-giving energy from the earth. The symbolism of the heart (for westerners the seat of the soul, the emotions, the capacity for love, and ultimately, life), the eyes (the windows to the soul for many cultures), and the brain (the locus of identity – and recently of life – for both the west and many tribal cultures), are central to many patients' experience of health and illness. The blood, hands, ears, nose, mouth, hair, breasts, liver, spleen, and bones all have important symbolic meanings that vary with culture, which if overlooked or misinterpreted may have serious consequences for diagnosis and treatment.

Beyond the body's symbolic meanings are the difficulties that can arise from conflicting cultural depictions of the interrelatedness of the body's parts. Although western medicine has paid lip service to holistic interpretations of health, scientific diagnosis and treatment rely on a mechanistic view that divides the body into distinct pieces and functions. The Asian practices of diagnosis via the pulses and treatment with acupuncture are unintelligible in the west, as they assume associations among organs and regions that western physicians believe to be totally separate [36]. In addition to formal cultural perspectives on anatomy and physiology, patients harbor remarkable beliefs about the size, number, location, functions, and behavior of what lies beneath their skin, and interpret their experiences accordingly [33].

Physicians who explain to patients about their medical diagnoses and the mechanisms of proposed interventions can help patients comprehend and accept treatment [47]. However, when the patient and physician do not have a common view of the nature of health and illness, the causes of particular conditions, and the meaning of certain ailments, "factual" scientific information is unlikely to have much positive effect on the patient's understanding or behavior [48]. Thus western efforts to educate Africans about the sexual transmission of AIDS have been complicated by the African cultural heritage that stresses the links between social harmony, fertility, and health. In a culture that equates infertility with death, HIV-infected women may continue to have children rather than lose their communal identity [67]. And where all living things – and many inanimate objects – are held to have spirits, even someone who accepts germ theory may assume that HIV chooses to infect certain persons and not others [35].

THE ROLES OF THE PATIENT AND HEALER

Closely related to the definition of disease is the deceptively complex issue of identifying the patient. Despite the truism that "the patient is the one with the disease," different cultural conceptions of disease may describe a number of "patients," each involved differently with the causes, symptoms, and treatment of a given condition. The western focus on the patient as an individual – for science, the individual with physical symptoms, for ethics the individual with decision-making authority – routinely conflicts with other cultural definitions that recognize the broader social effects of illness and group responsibility for the restoration of health.

In cultures where group links and social relationships are essential to identity, few conditions are perceived to affect persons as distinct individuals [16]. Even biological disorders in one body may be related to "disease" in another. In traditional Mexican belief, an infant's fallen fontanelle can be a sign of disease in the mother: the baby's fontanelle may collapse from the increased suction necessary to nurse from a mother with a poor flow of milk. Whereas scientific medicine would treat the infant for dehydration, traditional therapy would focus on rehydrating the mother as well.

Nonwestern views of family identity are particularly important in situations where health problems might affect procreative abilities. Where arranged marriages are common, the health and fertility of each of the family members are essential to their collective hopes of finding mates and their future status within the community [64]. Not only might the young man with "amenorrhea" fear an isolated old age without the support of children, the threat of his sterility may imply the death of his entire family line. In some traditional cultures a child with birth defects or a particularly stigmatizing condition may be less a patient than a symptom of the family's "disease"; in some Chinese settings, relatives may conceal an invalid's health problem, even from medical personnel, because of its potential social repercussions for the family [40].

Family-centered conceptions of the sick role naturally call for family-centered responses to disease. In many nonwestern systems, the family elders – typically male in most Hispanic, Asian, and Middle Eastern cultures [27, 57, 62], typically female in most African cultures [35] – assess the patient's condition and its ramifications, determine the most appropriate healer to consult, and interact with the practitioner to ensure the most appropriate treatment. Family members assume the bulk of the patient's personal care, provide a constant, supporting presence, and facilitate the patient's cooperation with the practitioner [27]. The patient's primary responsibility is to trust them to act in his or her best interests as a member of the family.

The importance of group relationships and interpersonal trust also shape the anticipated role of the physician and other caregivers in the traditional cultures of Asia, Latin America, India, the Middle East, and much of Africa. Medical professionals are not simply factual authorities whose knowledge and technical skill gives them the ability to help others; they are the community's moral experts, whose education, experience with human suffering, social position, and healing powers give them the moral authority to act on behalf of others [24, 38, 53, 55, 57]. In these cultures, the practitioner is expected to have the wisdom and the courage to make and carry out difficult decisions on the patient's behalf.

This view of the physician as paternalistic decision maker closely resembles the very tradition that modern medical ethics has challenged so successfully in the west. Yet, attempts to establish a patient-centered partnership between western physicians and patients from nonwestern cultures risk destroying any positive relationship that may exist. While patients almost universally appreciate practitioners who give them information and reassurance, the physician who asks the nonwestern patient's opinion of a treatment plan often *loses* the patient's trust and respect [24, 53, 55, 57]. In such situations, asking for a layperson's perspective is not a sign of the doctor's consideration for the patient's or family's goals and values, but rather an indication of the professional's weakness of character and lack of faith in the recommended treatment. Many non-westerners find the western practice of informed consent to be a cruel burden on the patient, who they believe should be relieved of such decision-making responsibilities as a consequence of being ill.

Moreover, to patients from such cultures, the written consent form, which is intended to clarify and document communication about the patient's care, is evidence of bad faith on the part of the doctor. Throughout the nonwestern world, where binding oral agreements are the norm in all but the most complex situations, consent to treatment is implied in the very act of seeking care; documentation of the patient's consent and the physician's beneficent intent is unnecessary [56, 57, 62]. Many patients unaccustomed to the western reliance on written contracts are terrified by the thought of what they might be signing away, as formal documentation portends significant risks. Illiterate patients and patients who do not speak English are particularly vulnerable to such an interpretation, as they are neither able to understand the purpose of the foreboding document nor its language. Some may refuse treatment rather than sign a document whose nature they do not understand [58] (Heitman, unpubl.).

The application of western standards of informed consent has posed ethical conflict in international AIDS research [4, 5, 11, 34, 46]. Clinical research with human subjects has its own complications for disclosure and consent: the risks of experimental intervention are potentially much greater than are those of established therapy, and the subject's best interests may be subordinated to the scientific rigor of the protocol. It is widely acknowledged that to avoid inflicting injustice on individuals suffering from disease, research subjects' consent to participation must be obtained regardless of their cultural background. Nonetheless, western ethicists have anguished over the implicit consequence that often no treatment will be available for AIDS patients who, due to their cultural conceptions of personal moral authority, cannot give individual consent to participate in a therapeutic drug trial [46].

REFLECTIONS ON DIALOGUE IN INTERCULTURAL MEDICAL ETHICS

The ethical complexity of international AIDS research has highlighted two challenging paradoxes in intercultural medical care: How can western doctors provide culturally appropriate treatment without sacrificing their own commitments to both their patients' health and the scientific principles of western medicine? How can western practitioners dedicated to the principle of individual autonomy best respect individuals who may not believe themselves to be autonomous?

These twin dilemmas echo another question born of cultural pluralism that should be familiar to every religious scholar: How can one affirm the validity of other religions while remaining faithful to personal views of religious truth? Just as increased contact and conflict among the world's religions has prompted clergy and academic religionists to establish a multifaceted, international dialogue on the nature of religious truth, medical ethicists and practitioners must look to intercultural dialogue to provide insight into medical and medical-ethical truth. The image of interreligious dialogue as an open-ended, ongoing process of mutual inquiry and self-discovery, in which participants explore their most sacred convictions and practices, can serve as a valuable model for clinical interaction in a pluralistic world.

The principles of interreligious dialogue are well suited to the vision of respect for persons upon which western medical ethics is based and the open-minded inquiry essential to good science [42]. Dialogue begins with a respect for others' views of themselves, as well as an eagerness to share one's own. Dialogue is not about preaching and conversion, but about mutual teaching and learning. In the universal search for truth, whether religious or scientific, no one can legitimately claim to have the single final answer. It can be enlightening for all to compare the truths that each possesses, and revealing to hear others' interpretations of one's cherished beliefs and unconscious assumptions.

Dialogue must rest on self-criticism and a willingness to hear and consider others' conceptions of their own systems. Although proponents of certain views may hope to convince others of their superiority or offer correctives to a one-dimensional interpretation, a truly open exchange may instead reveal the inconsistencies and incompleteness of established beliefs on all sides [42, 70]. Such revelation is often unsettling, but may afford new intellectual and moral insights and lead to new approaches to apparently irresolvable conflict. New questions that arise from such interaction can provide opportunities to bridge the gaps between disparate beliefs and peoples.

Several of the topics already under formal discussion among proponents of the world's religions have important ramifications for intercultural dialogue in medicine and medical ethics. The definition and source of life and the causes and meaning of death – essential to much of medicine's work – are fundamental religious questions, whose answers vary with religious tradition. Perhaps most important is the nature and logic of human suffering, including the relationship between human action and ultimate structures of justice. Where disease and disability are believed to be forms of suffering that result from divine punishment or *karma*, patients may respond quite differently to infirmity than would those who conceive of illness as an amoral phenomenon [32]. Medical care, too, takes on an important moral character when the physician's role is understood to be that of an instrument of divine mercy or compassion, religious concepts whose definition is a theme in dialogue among Christians and Buddhists.

Where the relationship between personal faith and religious authority is an important theme in interreligious dialogue, the tension between individual autonomy and social hierarchy are a rich terrain for intercultural dialogue in medical ethics. Western ethicists and social scientists alike have interpreted the group-oriented response of nonwestern cultures to illness in light of the dominant themes in western ethics and sociopolitical theory: self-determination and social justice. Commentators on western standards of informed consent in international research have analyzed the issue in terms of the use and abuse of personal and institutional power, and whether "trusted community leaders" can ever possess the moral authority to consent to participation for more vulnerable others.

Considered in dialogue with their counterparts from Hispanic, African,

Asian, and Middle Eastern societies, however, new issues emerge that would reshape the debate on patient autonomy. What western analyses do not explore, and what is much more interesting for meaningful intercultural policy on consent, is that in these same hierarchical cultures, even persons of considerable status lose their authority when they become ill. The head of an extended household would be expected to give up the mantle of authority as a patient; decisions about treatment would be made by others. Appropriate treatment would still be defined in terms of the patient as a member of the family or community, as his or her best interests could not be separated from those of the larger group. Although concepts of social hierarchy are important, the dominant theme in medical decision making in such cultures is responsibility for vulnerable others and group well-being.

A nonwestern perspective on western ideals of the patient's role would likely reinforce the clinical observation that many patients are abandoned to autonomy without the human support that can make self-reliance less onerous [37]. Western democratic views of the oppression and manipulation of powerless persons in hierarchical societies could be countered by criticism of the mistrust and indifference to personal limits evidenced in the western reliance on written consent, and of the apparent fragility of even the most important family relationships when illness strikes. Taken together, the ideals and criticisms of western and nonwestern medical systems converge on a central tension in human interaction: the relationship of trust, caring, and justice.

Dialogue on the ethics of human caring would not only serve medicine well, but would complement interreligious dialogue on global relationships. Dialogue presumes that, despite our differences, we ultimately live together, and that identifying and respecting mutual commitments is essential to peaceful coexistence. Many religious scholars stress the value of working together on common problems as a way to learn about others and oneself, and to build a new communal ethic from common experience [43, 70]. Like religious truth, both medical and ethical truth are revealed in their application. The universal threat of human disease and disability has long been a problem that fosters broader alliances among those who struggle jointly to understand and resolve it.

The lessons of dialogue are also taught by example. Much of western ethicists' impatience with other cultures' paternalistic medicine stems from the fact that the west has just recently rejected its own, ancient ethics of paternalism. The animated discussion among ethicists and clinicians from western democracies about the degree of autonomy that patients desire, which decisions they should make for themselves, and how best to respect their wishes [22], illustrates that the process is not complete. If, as some western theorists argue, individuation and respect for autonomy is the most evolutionarily advanced form of morality, the west's example should lead others to an appreciation and self-realization of autonomy in new cultural contexts. If autonomy is not a moral drive inherent in all persons, or if, as other western theorists contend, relational responsibility is the fuller expression of morality, the examples of nonwestern group interaction may enhance western appreciation of its worth.

As with interreligious dialogue, intercultural dialogue in medicine and medical ethics must take place on two levels: that of structured interaction and analysis among theoreticians and official "experts," and in the clinical encounters of individual patients and caregivers. Each ultimately must inform the other [43]. In each, openness and good will toward others, and solid knowledge of one's own and others' beliefs, are essential. While the dialogue appears to be underway at both levels [8, 15, 44, 69], and new international organizations have dedicated themselves to the study and discussion of intercultural medical ethics,[5] much remains to be done to provide a solid foundation of factual knowledge, both theoretical and applied. Clinicians who truly listen to patients will learn readily about alternative beliefs and practices [51, 64]. Both ethicists and practitioners would benefit as well from greater familiarity with the work of social scientists, and from more formal collaboration with them.

For practitioners, the model of the doctor-patient relationship as intercultural dialogue may seem unwieldy in clinical practice, where external factors limit the typical encounter to only a few minutes. However, such dialogue requires no more interaction than is called for by proponents of the medical covenant – whose ideal is an "I-thou" relationship of pure communication, or the medical contract – whose explicit disclosures and negotiation by both doctor and patient may be quite long and detailed. Additionally, it recognizes more overtly than either covenant or contract the limits of knowledge of self and other, and it offers practical gains for practitioners and patients who press those limits.

While the ethical goal of knowing each patient as an individual, and providing for his or her best interests, suggests that every patient must be more than part of an endless process of dialogue, gaining cultural insight from patients is not unlike gaining other clinical knowledge from their care. Throughout the history of medical practice, patients have always been part of the process of medical advance, a laboratory for the study of disease as well as, occasionally, the unfortunate victims of well-intended medical error. An appreciation of the therapeutic relationship as dialogue could restore an essential awareness of medicine's tragic dimensions to contemporary clinical ethics by highlighting the incompleteness and uncertainty of all medical care.

The common understanding of the western ethical mandate of informed consent would also benefit significantly from the insights of intercultural dialogue. The shortcomings of the typical real-life process of consent are demonstrated in its application in intercultural settings, where patients may be quick to sense that its focus, the consent document, is more for the physician's protection than for their own benefit. There may be no way to gain written consent from patients from oral cultures or nonliterate societies without risking coercing their signatures or destroying their confidence in the physician and the medical system. Explaining the consent form to patients unaccustomed to written contracts in health care should force clinicians and ethicists to re-examine the ethical commitments that create the therapeutic relationship, and how the consent document as it is currently used may undermine the mutual trust of patients and caregivers from all cultures.

Several existing interpretations of informed consent do apply to the practical issue of treating patients from cultures unfamiliar with the western view of autonomy and patient self-determination. By the "transparency standard" [6], adequate informed consent requires that physicians make their intentions and reasoning transparent to patients, out loud, rather than in writing, in language that the patient understands. Doctors must, first, provide information on proposed treatments and why they are preferable to other alternatives, and second, afford patients the opportunity to discuss this information to their satisfaction in a way that encourages them to make that information personally meaningful. Rather than relying on a stock disclosure for all patients, the transparency standard requires that the individual patient's needs and responses direct both the content and the process of the exchange.

The image of informed consent as a process of "negotiating clinical trust" [38] recognizes the ritualized dimensions of all doctor-patient interaction, and the need to clarify the symbolic, sociopolitical nature of this dialogue. Consent requires the doctor and patient to create a shared framework for understanding the context of the information, as well as the disclosure and questions themselves. Especially when the patient and physician speak different languages, but whenever their cultural backgrounds differ significantly, an interpreter or cultural mediator must assist each party in coming to know the other's intent. The interpreter's role as a cultural broker is particularly important in making the physician's thinking transparent to the patient through the lens of a different worldview. In keeping with consent theory, this description recognizes that trust is not and should not be given blindly, but must be created and sustained actively.

Consent as the "sharing of uncertainty" [29] addresses the underlying challenges of illness that all cultures and medical systems face: the experience of human suffering, the limits of human knowledge, and the unpredictability of the future. One of the most therapeutic effects of doctor-patient interaction results when doctors give patients the information and framework with which to integrate the experience of illness and healing into their lives [2, 25, 29, 37]. Such a process of informed consent engages, explores, and enriches patients' own explanatory models of illness and their ability to respond to further changes. By understanding the origins of patients' longing for certainty, and empathizing with their hopes, fears, and unrealistic wishes as they struggle to come to terms with illness, physicians can help their patients achieve psychological autonomy without imposing a western philosophical view of self-determination on nonwesterners. Such a process, moreover, may give insight into the roles that patients wish to take in their treatment, regardless of their cultural backgrounds, in a way that offers support as well as self-determination.

CONCLUSION

The past century's technological revolution has been marked by the transformation of once isolated and self-contained societies into an increasingly uniform, global culture where western science and material goods define much of human activity [61]. Western medicine is an essential dimension of this expanding technological culture, largely due to its tremendous successes. Yet, as medical technology reaches peoples unacquainted with western thinking, it promises them new moral questions as well as new medical benefits. Without a clear sense of dialogical nature of ethics and medicine, many in the west may be tempted to proffer established western analysis of these ethical issues as part of the package; they may be surprised when neither the technology nor the ethical theory is used as directed.

Commenting on the pitfalls of Christian missionary efforts, Ernst Troeltsch once stated that "the great religions might indeed be described as the crystallization of the thought of great races. There can be no conversion or transformation of one into the other, but only a measure of agreement and mutual understanding" [68]. Troeltsch's remarks are applicable to the missionary efforts of contemporary western medicine, and pose an ongoing challenge to medical ethicists. Ultimately no single vision of medicine or ethics will be adequate for the world's needs and true to the totality of human experience. Intercultural dialogue can enrich all of these varied interpretations, and the lives of those whom professionals in medicine and ethics aspire to serve.

The University of Texas School of Public Health
Program on Humanities and Technology in Health Care
Houston, Texas
U.S.A.

NOTES

[1] One of the challenges for any attempt to teach or learn about the medical practices and ethical beliefs of other cultures is the difficulty of identifying a group or individual without invoking labels or stereotypes. Even the designations that different cultural groups are given or use themselves vary with language, society, nation, and generation, often with important political significance.

In Great Britain, for example, "Asian" typically refers to people of Indian, Pakistani, and Bangladeshi descent, whereas in the United States it generally refers to people of Chinese, Japanese, Korean, Filipino, Vietnamese, and southeast Asian heritage. In the past few years, while Americans of African descent have debated whether to call themselves "black" or "African-American," non-African people of color in other parts of the English-speaking world have adopted the designation "black" to define themselves in contrast to whites, who may themselves be of mixed ethnic heritage. In either case, the terms lump together peoples of many highly diverse cultures, ostensibly because they have more in common than not, or because what they share is more important than what distinguishes them from each other.

The classifications used here – particularly the common terms western, nonwestern, African, Arabic, Asian, Hispanic, and North American – are intended to be general and somewhat fluid, recognizing that there are myriad cultures within each of these broader categories, and that individuals may not subscribe to the definition that an anthropologist or statistician would use to describe them [30, 49, 61]. Generalizations are helpful insofar as they facilitate initial introductions; but because ethical health care must be tailored to the patient, it is essential to recognize the limits of generalizations, and to understand individuals' beliefs and behavior in light of their own personal interpretations.

2 The past two decades have also seen tremendous immigration to Canada and western Europe, and its effects on their various medical systems have been similar to those described here. European health care professionals may be somewhat more perplexed by patients of different cultural backgrounds than are their North American counterparts, as European national identities traditionally have not admitted much ethnic diversity.

3 Additionally, major medical centers around the world often attract large numbers of foreign patients, who come seeking treatment unavailable in their own countries. Unlike many resident immigrants, these patients and their families are typically well off financially, and are accustomed to a certain degree of power within their own societies. As many such patients require long-term treatment or follow-up – often at great physical, emotional, and financial costs – they too may pose special challenges for caregivers (Heitman, unpubl.).

4 Evil eye is a world-wide phenomenon, particularly common in Hispanic, Mediterranean, and African cultures. Its etiology varies somewhat with locale, but the condition is typically considered to result from the prolonged stare or random compliment of someone who is envious of another's health, beauty, or success [27, 33]. Victims are typically young children, especially babies, and pretty young women, whose physical appeal prompts others' jealousy. Among Hispanics, the spell can be broken if the admirer touches the child while complimenting his or her appearance. Worldwide, many people wear amulets and charms to ward off the evil eye; western pediatricians may upset mothers greatly with warnings that such necklaces pose a threat to the child's health and safety.

5 These include particularly the European Society of Philosophy of Medicine and Health Care, the International Society of Bioethics, the International Society of Technology Assessment in Health Care, and the Society for Bioethics Consultation.

BIBLIOGRAPHY

1. Aday, L. A.: 1992, *Health and Health Care of Vulnerable Populations*, Jossey-Bass, Inc., San Francisco, CA.
2. Anderson, H.: 1989, 'After the Diagnosis: An Operational Theology for the Terminally Ill', *Journal of Pastoral Care* 43, 141–150.
3. Angel, R., and Thoits, P.: 1987, 'The Impact of Culture on the Cognitive Structure of Illness', *Culture, Medicine, and Psychiatry* 11, 465–494.
4. Angell, M.: 1988, 'Ethical Imperialism? Ethics in International Collaborative Clinical Research', *New England Journal of Medicine* 319, 1081–1083.
5. Barry, M.: 1988, 'Ethical Considerations of Human Investigation in Developing Countries: The AIDS Dilemma', *New England Journal of Medicine* 319, 1083–1086.
6. Brody, H.: 1989, 'Transparency: Informed Consent in Primary Care', *Hastings Center Report* 19 (September/October), 5–9.
7. Bochner, S.: 1983, 'Doctors, Patients, and their Cultures', in D. Pendelton, and J. Hasler (eds.), *Doctor-patient Communication*, Academic Press, London, pp. 127–138.
8. Carmody, D. L., and Carmody, J. T.: 1988, *How to Live Well: Ethics in the World Religions*. Wadsworth Publishing Co., Belmont, CA.
9. Cassell, E.: 1985, *Talking with Patients*, vols. 1 & 2, MIT Press, Cambridge, MA.

10. Chang, B.: 1981, 'Asian-American Patient Care', in G. Henderson, and M. Primeaux (eds.), *Transcultural Health Care*, Addison-Wesley Publishing Co, Reading, MA, pp. 255–278.
11. Christakis, N. A., and Panner, M. J.: 1991, 'Existing Ethical Guidelines for Human Subjects Research: Some Open Questions', *Law Medicine and Health Care* 19, 214–221.
12. Clements, C. D.: 1983. 'The Bureau of Bioethics: Form Without Content is Meaningless', *Perspectives on Biology and Medicine* 27, 171–182.
13. Conrad, P.: 1987, 'The Noncompliant Patient in Search of Autonomy', *Hastings Center Report* 17 (August), 15–17.
14. Crigger, B.-J., Campbell, C. S., and Homer, P.: 1989. 'Medicine, Morality, and Culture: International Bioethics, *Hastings Center Report*, Special Supplement (July/August), 1–32.
15. Cousins, N.: 1979, *Anatomy of an Illness as Perceived by the Patient*, W. W. Norton & Co., New York.
16. de Craemer, W.: 1983, 'A Cross-cultural Perspective on Personhood', *Milbank Memorial Fund Quarterly/Health and Society* 61, 19–34.
17. Dickens, B. M., Gostin, L., and Levine, R. J. (eds.): 1991, 'Research on Human Populations: National and International Ethical Guidelines', *Law, Medicine & Health Care* 19, 157–295.
18. di Matteo, M. R.: 1979, 'A Social-psychological Analysis of Physician-patient Rapport: Toward a Science of the Art of Medicine', *Journal of Social Issues* 35, 12–33.
19. Dvorak, E. M., and Waymack, M. I.: 1992, 'Is it ethical to recruit foreign nurses?', *Nursing Outlook* 39 (May/June), 120–123.
20. Elder, A. and Samuel, O.: 1987, *'While I'm Here, Doctor': A Study of the Doctor-patient Relationship*, Tavistock Publications, London.
21. Eliade, M.: 1959, *The Sacred and the Profane: The Nature of Religion. The Significance of Religious Myth, Symbolism, and Ritual within Life and Culture*, Harcourt Brace & World, New York.
22. Ende, J., Kazis, L., Ash, A., and Moskowits, M. A.: 1989, 'Measuring Patients' Desire for Autonomy: Decision Making and Information-seeking Preferences among Medical Patients', *Journal of General Internal Medicine* 4 (January/February), 23–30.
23. Engel, G. L.: 1977, 'The Need for a New Medical Model: A Challenge for Biomedicine', *Science* 196, 129–135.
24. Fitzgerald, F. T.: 1988, 'Patients from Other Cultures: How They View You, Themselves, and Disease', *Consultant* 28(3), 65–77.
25. Frankl, V. E.: *The Doctor and the Soul: From Psychotherapy to Logotherapy*, Vintage Books, New York, 1986.
26. Gahr, E.: 1992, 'Foreign Country Doctors', *Insight* 2 (May), 6–11, 37.
27. Galanti, G.-A.: 1991, *Caring for Patients from Different Cultures: Case Studies from American Hospitals*, University of Pennsylvania Press, Philadelphia.
28. Gorlin, R., and Zucker, H. D.: 1983, 'Physicians' Reactions to Patients: A Key to Teaching Humanistic Medicine', *New England Journal of Medicine* 308, 1059–1063.
29. Gutheil, T. G., Bursztajn, H., and Brodsky, A.: 1984, 'Malpractice Prevention through the Sharing of Uncertainty: Informed Consent and the Therapeutic Alliance', *New England Journal of Medicine* 311, 49–51.
30. Hanson, F. A.: 1975, *Meaning in Culture*, Routledge & Kegan Paul, London.
31. Hausman, W., Garrard, J., and Hong, K. M.: 1983, 'National Culture vs Professional Culture: A Comparative Study of Psychiatric Educators in Two Countries', *Culture, Medicine, and Psychiatry* 7, 23–34.
32. Heitman, E.: 1992, 'The Influence of Values and Culture in Responses to Suffering', in P. L. Starck, and J. McGovern (eds.), *The Hidden Dimension of Illness: Human Suffering*, National League for Nursing Press, New York, pp. 81–103.
33. Helman, C. G.: 1990, *Culture, Health and Illness*, 2nd ed., Wright, London.
34. IJsselmuiden, C. B., and Faden R.: 1992, 'Research and Informed Consent in Africa – Another Look', *New England Journal of Medicine* 326, 830–834.

35. Janzen, J. M.: 1978, *The Quest for Therapy: Medical Pluralism in Western Zaire*, University of California Press, Berkeley, CA.
36. Kaptchuk, T. J.: 1983, *The Web that Has No Weaver: Understanding Chinese Medicine*, Congdon & Weed, New York.
37. Katz, J.: 1984, *The Silent World of Doctor and Patient*, Free Press, New York.
38. Kaufert, J. M., and O'Neil, J. D.: 1990, 'Biomedical Rituals and Informed Consent: Native Canadians and the Negotiation of Clinical Trust', in G. Weisz (ed.), *Social Science Perspectives on Medical Ethics*, University of Pennsylvania Press, Philadelphia, pp. 41–63.
39. Kilborn, P. T.: 1991, 'Foreign-born Doctors Attracted to Rural Areas', *The Houston Chronicle* November 10, p. 11A.
40. Kleinman, A.: 1980, *Patients and Healers in the Context of Culture*, University of California Press, Berkeley, CA.
41. Kraut, A. M.: 1990, 'Healers and Strangers: Immigrant Attitudes Toward the Physician in America – A Relationship in Historical Perspective', *Journal of the American Medical Association* 263, 1807–1811.
42. Kung, H., van Ess, J., von Stietencron, H., and Bechert, H.: 1986, *Christianity and the World Religions: Paths of Dialogue with Islam, Hinduism, and Buddhism*, P. Heinegg (trans.), Doubleday & Co., Inc., Garden City, NY.
43. Kung, H.: 1991, *Global Responsibility: In Search of a New World Ethic*, Crossroad, New York.
44. Larue, G. A.: 1985, *Euthanasia and Religion: A Survey of the Attitudes of World Religions to the Right-to-Die*, Hemlock Society, Los Angeles, CA.
45. Lazare, A.: 1987, Shame and humiliation in the medical encounter. *Archives of Internal Medicine* 147, 1653–1658.
46. Levine, R. J.: 1991, 'Informed Consent: Some Challenges to the Universal Validity of the Western Model', *Law, Medicine & Health Care* 19, 207–213.
47. Ley, P.: 1988, *Communicating with Patients: Improving Communication, Satisfaction and Compliance*, Croom Helm, London.
48. Maclachlan, J. M.: 1958, 'Cultural Factors in Health and Disease', in E. G. Jaco (ed.), *Patients, Physicians and Illness: A Sourcebook in Behavioral Science and Health*, Free Press, New York, pp. 94–105.
49. Mbiti, J. S.: 1970, *African Religions and Philosophies*, Doubleday Anchor, Garden City, NY.
50. Mechanic, D.: 1972, 'Social Psychological Factors Affecting the Presentation of Bodily Complaints', *New England Journal of Medicine* 286, 1132–1139.
51. Murray, R. H., and Rubel, A. J.: 1992, 'Physicians and Healers – Unwitting Partners in Health Care', *New England Journal of Medicine* 326, 61–64.
52. Nursing Times Staff: 1981, 'Nurses "Misled" by Wrong Facts about Ethnic Communities', *Nursing Times* (April) 9, p. 624.
53. Ohnuki-Tierny, E.: 1986, 'Cultural Transformations of Biomedicine in Japan – Hospitalization in Contemporary Japan', *International Journal of Technology Assessment in Health Care* 2, 231–241.
54. Payer, L.: 1988, *Medicine and Culture: Varieties of Treatment in the United States, England, West Germany, and France*, Penguin, New York.
55. Pearson, R.: 1982, 'Understanding the Vietnamese in Britain. Part III: Health Beliefs, Birth and Child Care', *Health Visitor* 10, 533–540.
56. Perry, S. (ed.): 1986, 'Treatment of the Terminally and Critically Ill in Selected Countries (Abstract of a Panel Discussion)', *International Journal of Technology Assessment in Health Care* 2, 720–724.
57. Qureshi, B.: 1989, *Transcultural Medicine: Dealing with Patients from Different Cultures*, Kluwer Academic Publishers, London.
58. Qureshi, B.: 1990, 'Asian Patients: Bridging Cultural Gaps', *MIMS Magazine* (June 15), 25–27.

59. Reiser, S. J.: 1978, 'The Decline of the Clinical Dialogue', *Journal of Medicine and Philosophy* 3, 305–313.
60. Romanucci-Ross, L., Moerman, D. E., and Tancredi, L. R. (eds.): 1991, *The Anthropology of Medicine: From Culture to Method*, 2nd ed., Bergin & Garvey, New York.
61. Roosens, E. E. (1989). *Creating Ethnicity: The Process of Ethnogenesis*, Sage Publications, Newbury Park, CA.
62. Sampson, C.: 1982, *The Neglected Ethic: Religious and Cultural Factors in the Care of Patients*, McGraw-Hill Book Co. (UK) Ltd., London.
63. Sandler, R. H., and Jones, T. C. (eds.): 1987, *Medical Care of Refugees*. Oxford University Press, New York.
64. Scrivens, E., and Hillier, S. M.: 1982, 'Ethnicity, Health and Health Care', in D. L. Patrick, and Graham Scambler (eds.), *Sociology as Applied to Medicine*, Bailliere Tindall, London.
65. Siegel, B.: 1984, *Love, Medicine, and Miracles: Lessons Learned about Self-healing from Experience with Exceptional Patients*, Harper & Row, New York.
66. Tanaka-Mitsumi, J., and Higginbotham, H. N.: 1989, 'Behavioral Approaches to Counseling across Cultures', in P. O. Peterson, J. G. Draguns, W. J. Lonner, and J. E. Trimble (eds.), *Counseling Across Cultures*, 3rd ed., University of Hawaii Press, Honolulu, HI, pp. 269–298.
67. Tofani, L: 1991, 'Ancient Duties Overrule Health for Some Victims', *The Houston Chronicle* May 12, p. 22A.
68. Troeltsch, E.: 1981, quoted in J. Hick and B. Hebblethwaite, *Christianity and Other Religions*, Fortress Press, Philadelphia, p. 28.
69. Veatch, R. M. (ed.): 1989, *Cross Cultural Perspectives in Medical Ethics: Readings*, Jones and Bartlett, Boston, MA.
70. Welch. S. D.: 1990, *A Feminist Ethics of Risk*, Fortress Press, Minneapolis, MN.
71. World Health Organization (WHO): 1978, *International Classification of Diseases*, 9th ed., WHO, Geneva.
72. Zaner, R. M.: 1988, *Ethics and the Clinical Encounter*, Prentice-Hall, Englewood Cliffs, NJ.
73. Zola, I. K.: 1973, 'Pathways to the Doctor – From Person to Patient', *Social Science and Medicine* 7, 677–689.

JONATHAN R. SANDE

THEOLOGY, ETHICS, AND CLINICAL ENCOUNTERS:
POSSIBILITIES FOR RECONCILIATION?

This essay is not intended to serve as a summary analysis of the entire volume. Rather it is an attempt, using the help of the preceding pages, to establish possible trajectories for further work at the interfaces of theology, ethics, and clinical encounters. A narrative account of a particular clinical situation will provide the backdrop for a discussion of concerns that emerge from the preceding essays. Areas for further investigation will be suggested.[1]

＊   ＊   ＊   ＊   ＊

I entered his room to say farewell. The room and its contents were an example of the "best" that intensive medical care in the United States has to offer: ventilatory support, central venous lines for cardiac and pulmonary monitoring, automated monitors for blood pressure, pulse oximetry, and cooling blanket, bags, poles, hooks and pumps for IV solutions, and so forth. Yet, despite all this, the room was unusually quiet. Most of the alarms usually heard from such equipment had been disarmed after the decision to withdraw life support. Looking in vain for some response from the man I had cared for during the last several weeks, I was aware of sound from only two sources – the ventilator, and the radio. "He never wanted to be on life support," the family told us. "When he was on the farm he always left the radio on for his cows. He thought it gave them some peace. Please leave it on for him," the family told us.

Except for the technology invading his body and surrounding both of us he and I were alone in the room. My thoughts wandered a bit. I wondered about the clinical situation, about this gentleman who had come to us with an advanced leukemia. Aware of the risks, together we had embarked on chemotherapy. We hoped to kill the cancer cells in his bone marrow, realizing that with this treatment we also ran the risk of destroying the normal bone marrow cells he needed to survive. A "blood count ritual" developed as his hospitalization progressed. Each morning his blood would be drawn, and we would await his blood counts from the lab. Initially they went down as the chemotherapeutic poisons we administered attacked both good cells and bad. After that we waited and hoped for the day when his bone marrow would recover and his counts would begin to rise. Days, then weeks passed, and his counts remained low. He continued to require transfusion support for his hemoglobin and platelets; worse yet, he began to develop fevers. Because our chemotherapy had destroyed the white blood cells which fight infection, we were forced to begin high doses of powerful antibiotics. In spite of this he deteriorated; with each passing day he had more and more

225

G.P. McKenny and J.R. Sande (eds.), Theological Analyses of the Clinical Encounter, 225–233.
© 1994 Kluwer Academic Publishers. Printed in the Netherlands.

difficulty breathing. However, chest X-rays, cultures, and blood tests for potential infectious agents revealed nothing unusual. More and more physicians became involved in his care as we searched for and wondered what, if anything, we were missing, and what more, if anything, we could do. Finally we began a medication designed to enhance the growth of his bone marrow, aware that this carried with it the chance of stimulating not only his normal cells but also a recurrence of his leukemia. It didn't help. His blood counts remained low; he remained febrile; he continued to need transfusions; his condition worsened. He was transferred to the intensive care unit. The number of physicians involved in his care expanded still further. Conversation with him about what to do in case things continued to get worse were ongoing; we agreed that if he should need it, and if we as his physicians felt there were still a chance for him to come back, we could intubate him and put him on a breathing machine. This, indeed, happened; his breathing became very labored and his oxygenation increasingly poor. A breathing tube was inserted, and he was placed on a ventilator. We hoped we could buy him a few more days so that his bone marrow would begin to return to normal function. It didn't. His condition continued to deteriorate, and the family asked us to withdraw support.

As I stood at his bedside my thoughts wandered further, to my relationship with him. As the weeks passed we had grown to know each other better, our mutual histories and the communities from which we came. This had occurred, for the most part, from shared conversation. We talked about human physiology, about the this and that of his disease process and our attempts to treat it. We talked about the anxiety of waiting for his blood counts to rise and the disappointment of seeing no improvement. We talked about his increasing frustration as things continued to deteriorate. But we also talked about ourselves, and some of our knowledge of each other came through more than shared conversation. It also came through shared experience and shared emotion. We talked of his fatigue, but he sensed and asked about the increasing signs of fatigue I displayed during the time we had together. He asked about my family and my life. And, as the end drew near, we spoke of children. He spoke of his children and grandchildren, and I told him of my children. We shared pictures of our families. We agreed that in many ways children are what life is all about. Especially early on we laughed and joked; this became more and more infrequent as time passed. The last laugh we shared was at my expense. During morning rounds the patient told the attending physician that he had appreciated the back rub I had given him during an episode of rigors the night before, but that he wished that someone more attractive had administered it.

Alone with him at the bedside, I also wondered about the numerous moral issues involved. Was he competent in the last hours before intubation? Did he really want us to go ahead, or was he simply frightened? "Do what you think is best," he said. "I trust you." But we and the family continued to wonder if we had done the right thing. We wondered about the length of time we should continue to support him, especially after it became necessary to sedate him because of his discomfort with the respirator. What would he have us do? But he could not tell us. And, toward the last, even though we still held a glimmer of hope for him, several physicians wondered amongst themselves about the extensive resources devoted to his care. By coincidence, the morning the family decided to withdraw support physicians from Europe had visited our hospital. We discussed ways in which our handling of this case differed from what might have been done in Europe. This served to increase our uncertainty about an already unsettling situation.

As I said my farewell to him I knew that in a short while he would be gone. I couldn't help but wonder, for a fleeting moment, where he would "go." He had wondered aloud about God's will for him, as did the family, especially near the end. I had wondered, silently, about the same question. The chaplain spent many hours with them during the last days. Slowly we all came to accept what appeared to be inevitable. With tears and sadness, we had, as a group, agreed to withdraw support. Each of us, for our own reasons, wanted to be alone with him over the last several hours; each of us, in our own way, found that time.

I held his hand, wondering about such things, talking to him, saying how much I wished things had turned out differently. He continued to be unresponsive. With tears in my eyes I held his hand for one last time, stroked his now bald head, and told him I would miss him. I noticed moisture seeping from his closed left eye. Was it a tear, or was it a result of ongoing baseline physiologic secretion? Perhaps, for one last time, we had touched each other.

*   *   *   *   *

The authors of each essay in this volume were asked to focus on one particular aspect of the interfaces between clinical encounters, ethics, and theology. In spite of these different particulars common themes appear in the preceding essays. One way to focus on these themes and their significance for further investigation is to reflect, in a general way, on clinical encounters, ethics, and theology, and to approach the essays with these reflections in mind.

Clinical encounters are complex events indeed. The word *clinical* is derived from the Greek *klinikos*, meaning pertaining to a bed [2]. Thus the adjective clinical places us at the bedside, from which we observe and respond to the course of disease in the patient. Clinical medicine, and the "clinical sciences" on which it is based, stand in contrast to the basic medical sciences. Examples of the basic medical sciences are the studies of physiology and pathophysiology in immunology, biochemistry, microbiology, and so forth, which most often take place at least one step removed from the bedside. Many commentators and most clinicians agree that clinical medicine is neither wholly science nor wholly art. Rather it is a combination of the two, or even something unique unto itself, that requires not only the knowledge of physiology and pathophysiology provided by the basic sciences, but also a combination of skills, judgment, experience, and intuition through which this basic knowledge is most appropriately used for the benefit of the patient.

That good judgment, experience, and intuition are necessary for excellence in clinical medicine is made clear by the following characteristics of clinical medicine (many more could be named): the difficulty of recognizing a general pattern of disease in a particular patient who may not "fit" classic diagnostic categories; the difficulty of knowing completely, much less calculating accurately, risks and benefits of various treatment modalities in a particular patient; and the lack of good outcome studies for many so called "well-accepted" treatment modalities which would make such calculations easier. As a result, the "science" to which some clinicians like to refer, and which medical students, interns, residents, and fellows spend so much time pursuing, is only a small part of a much larger story. In the end the practice of clinical medicine is marked by judgments of varying degrees of uncertainty. This uncertainty is most marked in the "tough" and "humbling" cases, but is present even in "common everyday" cases.

Uncertainties stemming from the nature of *clinical* medicine are often magnified by the nature of the human *encounter* between patient and doctor, a dimension of clinical encounters often neglected by clinical medicine and bioethics alike. Permeating each encounter between patient and doctor are basic characteristics of human existence, including the experiences of selves in time, sociality, and otherness. Patient and doctor encounter one another as human beings living in the present, yet each has a past manifest in the present by memories, life work, and learned patterns of response and habituation, and each has a future manifest in the present by anticipation, hope, and fear. Patient and doctor encounter one another as human beings deeply embedded within social matrices; each has or has had communities of formation and sustenance which most likely are quite different from each other; each also, in the encounter with one another, is drawn into mutual relation with common third parties, be it nursing staff, hospital administrators, third party payers, lawyers, the government, or even bioethicists. Patient and doctor encounter one another as human beings with more and less self-knowledge, at different levels of maturity and different capacities for growth and adaptation to circumstances, and different sensitivities to and experiences of other objects and persons; ultimately each experiences the other as one who is foreign and not self.[2] These basic human experiences of selves in time, sociality, and otherness influence the judgments made in clinical encounters – in varying degrees, to be sure, and in both positive and negative ways, to be sure. For example: a physician, remembering her last multiple myeloma patient, states "Because I treated Mr. B with this and he reacted in such and such a way I think I will treat Mr. C differently"; or because a patient does not have insurance a hospital elects to deny treatment over the objections of both patient and doctor; or a white male patient decides that because his physician is neither white nor male he will look for a different one.

On this reading not only is there room for error and miscalculation in clinical encounters because of the nature of clinical medicine (for example, simply not knowing enough, or missing a diagnosis, or miscalculating the risk of treatment in a particular patient); there is also room for error and miscalculation in clinical encounters because of the nature of the human encounter between doctor and patient (for example, not appreciating the extent to which the past experience of caring for a sister dying from leukemia informs a newly diagnosed lymphoma patient's refusal of possibly curative chemotherapy, or being convinced that patients will sue and thus ordering too many tests, or letting the offensiveness of a patient's morbid obesity stand in the way of not only appropriate medical therapy but also appropriate human relations with her). Indeed, a careful observer of physicians might well wonder which kind of error is more frequent and which more serious.

Parts of the preceding essays can be seen as examining various aspects of such difficulties. For example, May probes the relation between the intellectual activity of physicians (for example, clinical judgment) and the virtue of prudence and its three elements: *memoria* (being truly open to the past); *docilitas* (an openness to the present); and *solertia* (an openness to the future). DuBose enlists the aid of "postmodern" thought to argue that we ought to be suspicious of the ability of physicians to know the best interests of patients, to present facts to patients objectively, or to exhibit the virtues supposedly inherent in their professional roles. This is not necessarily because physicians are of bad character, but rather because of the difficulties of finding and establishing trust in this "postmodern age of suspicion." Sulmasy questions the view of human being (that is, the "anthropology") implied by sociological and political models undergirding current discussions of authority in patient doctor relationships, and argues for a revised understanding of authority based on scripture and on a different understanding of the nature of human being. Glaser points to the obvious but too often neglected fact that patient and doctor encounter one another in organizations which are in turn located within societies which are in turn part of a global community. He argues that individuals, organizations, and societies need to attend much more closely to these broader social dimensions of moral decision making. Lebacqz points to the inequity in and abuses of power that pervade clinical encounters. Zaner uses a phenomenological description of a clinical encounter to underscore both the reflexive relations present in clinical encounters and what he calls a fundamental moral sense evoked by the other. He argues for an ethic of "affiliative feeling" involving compassion, care, and trust. Heitman points out the numerous barriers to effective medical care that cultural differences create and argues for a solution incorporating dialogue with the other.

Ethics, most especially bioethics, has become a "growth industry." Over the last thirty years an enormous amount of literature has been published in bioethics, especially in the United States. Until recently concerns dominating the field could be subsumed neatly under the "mantra of bioethics," that is, the three ethical principles of respect for autonomy, beneficence, and justice. These principles have fostered much discussion of issues such as: maintaining respect for patient wishes, informed consent, end of life concerns including advanced directives and physician assisted suicide, and the appropriate allocation of scarce resources. If one assumes that the volume of literature published in a field correlates with success bioethics would be a huge success indeed.

However, there are reasons to wonder whether bioethics is as successful as it might appear. For one example, in spite of nearly three decades of literature arguing for the importance of patient empowerment and autonomy, it was only at a recent bioethics conference, according to Arthur Caplan, that families of patients involved in the famous "right to die" cases were featured as thoughtful people who could contribute to ongoing discussion of these issues by academic "experts" [1]. For another example, the place of ethics in relation to medicine is still strenuously debated. Positions taken span a range from those who believe bioethics to be an imposition by outsiders seeking to control medicine to those who believe ethics to be an intrinsic dimension of professional activity. (One need not presuppose that agreement on this vexed question is a necessary criterion of success in bioethics to be troubled by the apparent lack of consensus on this issue.)

Various critiques of the "success" of bioethics, including differing views of the relation of ethics to medicine, are apparent in the preceding essays. For example, Hamel finds in the covenant model a view of ethics as "corrective vision." He argues that the covenant model can correct and broaden understandings of the patient doctor relationship, and that it has the potential to contribute meaning to some basic, if troubling, human experiences. May asks what it means to be a physician, and brings forth ethical guidelines from his conclusions. Glaser develops an ethic based on beneficence within the conditions of finitude and argues that in our health care ethics decisions we must substantially broaden our field of concern. Some will see his ethic as a radical and transforming one, as could be expected from his adaptation of the Parable of the Good Samaritan. Lauritzen criticizes bioethics to date as too heavily tilted towards dry rationality, impartiality, and legalism, arguing that the attendant elimination of emotion as a relevant dimension of human moral experience both hinders moral insight and impoverishes moral discernment. He searches for a model of bioethics that will enable physicians to be intellectually *and* emotionally present to their patients. The essay by the Churchills argues for a transformation of our understanding of ethical theory. Their "collage" model turns away from a view of ethical theory as a source of unitary explanation of moral phenomena to one of ethical theory as enhancing interpretive power that would, in turn, enable more accurate "excavation" of clinical encounters and all that goes on in them.

By design each essay in this volume has a particular focus of its own. Yet when assembled together the individual foci – the nature of covenant relationships, the basis of professional activity and its attendant virtues, elements of clinical encounters that undermine trust, the modulation of autonomy, beneficence, and trust in clinical encounters provided by theology, the understanding of authority in clinical encounters, the complex institutional and social dimensions of ethical decisions in health care, the use of power in clinical encounters, the role of emotion in moral discernment, the appropriate role of ethical theory in moral discourse, the phenomenological basis of ethics in clinical encounters, and the problem of pluralism and cultural difference in clinical encounters – establish a trajectory somewhat different from that of the lions share of bioethics of the last three decades. This new trajectory invites us to use clinical encounters and ethics as a way to wonder about and explore deep and long standing human questions. These might be questions concerning the experience of patients with illness, or questions concerning the extraordinarily rich and powerful human relationship that is the encounter between patient and doctor. These might be questions concerning the nature of medicine as a healing activity, or questions concerning the nature of morality and ethical reflection.

The line of inquiry suggested by the last several pages, if developed further, might be summarized as follows. Clinical excellence requires more than mere knowledge of pathophysiology. It also requires the skills, judgment, experience, and intuition necessary to apply this knowledge to patients in the most appropriate fashion; and it requires knowledge of the nature of the human encounter between patient and doctor and ways in which this dimension of therapeutic relationships may influence outcomes. Similarly, excellence in bioethics requires more than facile use of principles. It also requires a kind of habituated sensitivity to the fine and variable contours of moral dilemmas, clear headed thinking about benefits and burdens, and prudence or moral discernment incorporating not only intellectual but also emotional rigor. It may well be that a fuller integration of the clinical and the ethical is needed, an integration in which both are full in themselves and yet together become more than either one alone, intertwined and mutually supportive. An endpoint of this line of reasoning is, of course, that ethics is not something separable from medicine and medical activity but is intrinsic to the nature of healing itself.

What role theology ought to play in bioethics is, like the relation of ethics to medicine, an issue on which there is much disagreement. Theologians are generally acknowledged to have played a prominent role in the development of bioethics in this country. But recently theological voices have tended to fade from the discourse of bioethics, or at least recede into the background as other voices have developed increased influence.

One might infer that the range of possible positions on this issue occupies a span between two poles. The position at one pole would say theology alone can and does ground ethics and the practice of medicine, whether or not this is acknowledged by non-adherents. The position at the other pole would say theology has nothing to contribute to ethics or the practice of medicine, whether or not this is acknowledged by adherents. While none of the essays in the volume occupy either extreme, some are clearly closer to one pole than the other. For example, Thomasma and Pellegrino, and Sulmasy, offer theological interpretations of ethics and the clinical encounter even while acknowledging powers of natural reason and what might be viewed as "secular equivalents" (though ultimately, in their eyes, less satisfactory "equivalents") to their theological interpretations. Thomasma and Pellegrino argue that theological backing enables a more profound and satisfying understanding of autonomy, beneficence, and trust in the encounter between patient and doctor than do secular equivalents; Sulmasy draws his understanding of authority as *exousia* (the moral warrant to heal ultimately granted by God) from theological warrants and the New Testament, again acknowledging possible but less satisfactory secular justifications for this "warrant to heal." Nearer the middle of the span are, for example, Heitman and Lauritzen, both of whom argue that theology can turn attention to dimensions of the lives of patients and doctors that otherwise would be neglected. On the other end of the span, DuBose raises the problem of sustaining simplistic claims to over-arching theological metanarratives in this "postmodern age of suspicion," though even he is not entirely pessimistic about the potential contributions of theology to bioethics. Far from resolving the question of the relation of theology to ethics and clinical encounters, then, the essays in this volume instead press forcefully the following question: Are there possibilities for reconciliation between theology and bioethics?

Reconciliation is derived from the Latin word *reconciliare*, which has been used to translate the Greek *katalassein*, which in turn has the root meaning of change. In theological and moral contexts reconciliation has come to be associated with conversion and forgiveness. It refers to a change in attitude, from one of hostility to one of friendship, not only of God toward humanity but also of humanity toward God and of individuals toward one another [5]. Obviously the term is relational in character. Yet it also points to a change of heart, a turning of the self, or a change in orientation of one being toward another, for example, of doctor toward patient.[3]

Even in this brief explication of reconciliation one can note parallels with some of the preceding essays. For example, May speaks of a professional as one who professes, which "means 'to testify on behalf of,' 'to stand for,' or 'to avow' something that defines one's fundamental commitment. A profession *opens out toward* an as yet unspecified transcendent good that defines the professor" (p. 33, emphasis mine). For another example, Sulmasy writes concerning his central concept of *exousia* that it is "both an orientation to virtue and the fruit of virtue" (p. 97), and that

to practice medicine with *exousia* is to ordain one's practice to the good of the patient and to ordain one's practice for the good of the patient to the glory of God. In this way, the *dynamis* to heal, which is already given in nature and in human reason, not only becomes actual but has a context and an ultimate orientation, emerging from God and leading back to God. *Exousia* therefore demands the virtues of practice: wisdom, equanimity, selflessness, trustworthiness, concern, and fidelity (p. 97).

In these quotes we see many of the concerns of the volume as a whole and this concluding essay in particular brought together. Reconciliation, or a change in heart or attitude, is brought near to a concern for proper orientation towards one's patient, one's profession, and ultimately toward God. According to these commentators, a physician stands for – or ought to stand for – some transcendent good. In turn the activity of physicians is a self-involving task demanding virtues not only related to the acquisition of knowledge, but also to the appropriate use of such knowledge in individual patients and a sensitivity to their humanity.

Thus reconciliation joins some of the central concerns of the essays of this volume. It speaks not only to what some clearly see as religious and theological dimensions of medical practice, but also to the fuller and richer understandings of clinical encounters and bioethics suggested above and in the preceding essays. Perhaps there are seeds to be nurtured here that may yet achieve some kind of reconciliation of theology, ethics, and clinical encounters that will mutually enrich all three – and, most importantly, serve our patients and God.

Mayo Clinic
Rochester, Minnesota
U.S.A.

NOTES

[1] A few words of explanation are in order. This essay is meant to be suggestive, in three senses. First, I will attempt to suggest some connections between the essays in the volume, and between concerns that I see emerging from the volume and trends in the bioethics literature. Second, these connections will be incompletely drawn. My hope is that these connections will be interesting enough – that is, suggestive enough – so that more work will be done in at least some of these areas. Third, the narrative account is intended to serve as a backdrop for the remainder of the essay; that is, I believe that the narrative account suggests certain things about the remainder of the essay, and visa versa. With more space these suggestions could be spelled out clearly; as it stands the reader is invited to wander back and forth between the narrative account and the corresponding parts of the essay.

[2] This paragraph owes much to H. Richard Niebuhr [3, 4], and to Richard M. Zaner [6].

[3] Of course, a development of H. Richard Niebuhr's thought concerning reconciliation would be very helpful here [3].

BIBLIOGRAPHY

1. Caplan, A.: 1992, 'Biomedical Ethics Captures Center Stage', an interview in *Minnesota Medicine* 75, 7–9.
2. *Dorland's Illustrated Medical Dictionary*, 1991, W. B. Saunders Company, Philadelphia, 278.
3. Niebuhr, H.: 1963, *The Responsible Self: An Essay in Christian Moral Philosophy*, Harper & Row, Publishers, New York.
4. Niebuhr, H.: 1989, *Faith on Earth: An Inquiry Into the Structure of Human Faith*, Richard R. Niebuhr (ed.), Yale University Press, New Haven.
5. O'Donovan, O.: 1986, *The Westminster Dictionary of Christian Ethics*, James F. Childress and John Macquarrie (eds.), Westminster Press, Philadelphia, p. 528.
6. Zaner, R: 1988, *Ethics and the Clinical Encounter*, Prentice-Hall, Englewood Cliffs.

NOTES ON CONTRIBUTORS

Larry R. Churchill is Professor and Chair of the Department of Social
Medicine, and Adjunct Professor in the Department of Religious Studies
at the University of North Carolina at Chapel Hill.

Sandra Wade Churchill is completing a Ph.D. at the Graduate School of the
Union Institute at Duke University.

Edwin R. DuBose is Senior Associate for Religious Ethics and Clinical Practice
at The Park Ridge Center for the Study of Health, Faith and Ethics in
Chicago.

John Glaser is Director of the Center for Healthcare Ethics for the St. Joseph
Health System in Orange, California.

Ron Hamel is Senior Associate for Theology, Ethics, and Clinical Practice
at The Park Ridge Center for the Study of Health, Faith and Ethics in
Chicago. He is also Co-Director of the Clinical Ethics and Medical
Humanities Program at the Lutheran General Health System in Park Ridge,
Illinois.

Elizabeth Heitman is Assistant Professor of Humanities and Technology in
Health Care at the University of Texas School of Public Health in Houston,
Texas. She is also Adjunct Assistant Professor of Religious Studies at Rice
University.

Paul Lauritzen is Associate Professor of Religious Studies at John Carroll
University.

Karen Lebacqz is Professor of Christian Ethics at the Pacific School of Religion
in the Graduate Theological Union, Berkeley.

Gerald P. McKenny is Assistant Professor of Religious Studies at Rice
University. He is also an adjunct faculty member of the University of
Texas Medical School at Houston and Co-Director of the Rice University
of Texas Joint Program in Health Care Ethics.

William F. May is Cary M. Maguire University Professor of Ethics at Southern
Methodist University.

Edmund D. Pellegrino is John Carroll Professor of Medicine and Medical
Ethics, Director of the Center for the Advanced Study of Ethics, and Director
of the Center for Clinical Bioethics at Georgetown University.

Jonathan R. Sande is a resident in Internal Medicine at the Mayo Clinic in
Rochester, Minnesota, and a doctoral candidate in theological ethics at the
University of Chicago Divinity School.

Daniel P. Sulmasy, a Fransiscan friar, is Assistant Professor of Medicine in
the Division of General Internal Medicine and a Research Scholar in the

235

G.P. McKenny and J.R. Sande (eds.), Theological Analyses of the Clinical Encounter, 235–236.

Center for Clinical Bioethics at the Georgetown University Medical Center.
He is also working toward a doctorate in philosophy.

David C. Thomasma is Michael I. English Professor of Medical Ethics and
Director of the Medical Humanities Program at Loyola University (Chicago)
Medical Center. He is also Chief of the Ethics Consult Service.

Richard M. Zaner is Ann Geddes Stahlmlan Professor of Medical Ethics,
Director of the Center for Clinical and Research Ethics, and Director of
the Clinical Ethics Consultation Service at Vanderbilt University Medical
Center.

INDEX

238 INDEX

Theology and Medicine

Managing Editor

Earl E. Shelp, *The Foundation for Interfaith Research & Ministry, Houston, Texas*

1. R.M. Green (ed.): *Religion and Sexual Health.* Ethical, Theological and Clinical Perspectives. 1992   ISBN 0-7923-1752-1
2. P.F. Camenisch (ed.): *Religious Methods and Resources in Bioethics.* 1994
                                                                ISBN 0-7923-2102-2
3. G.M. McKenney and J.R. Sande (eds.): *Theological Analyses of the Clinical Encounter.* 1994   ISBN 0-7923-2362-9

KLUWER ACADEMIC PUBLISHERS – DORDRECHT / BOSTON / LONDON

DATE DUE

			Printed in USA